Report of the
Defense Science Board Task Force
on
Mission Impact of Foreign Influence on DoD Software

September 2007

Office of the Under Secretary of Defense
For Acquisition, Technology, and Logistics
Washington, D.C. 20301-3140

The DSB is a Federal Advisory Committee established to provide independent advice to the Secretary of Defense. Statements, opinions, conclusions, and recommendations in this report do not necessarily represent the official position of the Department of Defense.

This DSB Task Force on Mission Impact of Foreign Influence on DoD Software completed its information gathering in November 2006.

This report is UNCLASSIFIED and is RELEASABLE to the public.

OFFICE OF THE SECRETARY OF DEFENSE
3140 DEFENSE PENTAGON
WASHINGTON, DC 20301-3140

DEFENSE SCIENCE
BOARD

September 21, 2007

MEMORANDUM FOR ACTING UNDER SECRETARY OF DEFENSE
(ACQUISITION, TECHNOLOGY, AND LOGISTICS)

SUBJECT: Final Report of the Defense Science Board Task Force on Mission Impact of Foreign Influence on DoD Software

I am pleased to forward the final report of the Defense Science Board Task Force on Mission Impact of Foreign Influence on DoD Software. This task force, chaired by Dr. Robert Lucky, was established to complement the 2005 DSB study on High Performance Microchip Supply, which focused on the implications of semiconductor fabrication in foreign countries. The task force found that the DoD faces similar consequences from the possible exploitation of software, increasingly developed outside the United States, in its systems.

The task force was asked to assess the Department's dependence on software of foreign origin and the risks involved. The task force considered issues with supply chain management; techniques and tools to mitigate adversarial threats; software assurance within current Defense programs; and assurance standards within industry, academia, and government.

The report addresses the future U.S. ability to ensure and maintain a trusted supply of software to DoD and the U.S. Government. In their report, the task force states that there is no absolute guarantee that software can be sanitized of all vulnerabilities, intended or unintended. The task force recommends a suite of processes and mitigation strategies to reduce the risk of interrupted system performance and ensure mission success.

I endorse the recommendations of the task force and encourage you to review their report.

Dr. William Schneider, Jr.

OFFICE OF THE SECRETARY OF DEFENSE
3140 DEFENSE PENTAGON
WASHINGTON, DC 20301-3140

DEFENSE SCIENCE BOARD

September 18, 2007

MEMORANDUM FOR THE CHAIRMAN, DEFENSE SCIENCE BOARD

SUBJECT: Final Report of the Defense Science Board Task Force on Mission Impact of Foreign Influence on DoD Software

The Defense Science Board (DSB) Task Force on Mission Impact of Foreign Influence on DoD Software has completed its work, and a final report is attached. The task force examined areas in software security, security architecture, and risk mitigation and received briefings from industry, academia, and a number of Defense agencies. Briefings on software assurance and development processes for Defense programs such as Blue Force Tracker, F-22, and Future Combat System were also provided to the task force.

The Department's dependence on software, which is growing in size and complexity, presents tempting opportunities for U.S. adversaries to exploit. Further, the increasing interconnectedness of defense systems could lead to the exploitation of many applications through a single vulnerability. These weaknesses, among others, are significant liabilities to the Department's mission-critical systems; however, DoD cannot ignore the economic advantage of globally-produced, commercial-off-the-shelf software. The globalization trend of the software industry will continue to occur, and some of DoD's software will be developed in foreign countries.

The task force found that low-level, malicious techniques have been employed to successfully penetrate sensitive, unclassified DoD systems despite efforts by DoD to maintain information security and assurance. DoD's current evaluation strategies and techniques are inadequate to deal with the growing functionality and outsourcing trend of software, making exploitation easier and defense more difficult. The problem is complex, and ultimately, an intelligent risk management process will be essential to ensure a trusted supply chain, mitigate malicious attacks, enable efficient responses and reactions, and maintain trustworthiness in the software that support DoD's critical missions.

The task force outlined 11 recommendations in this report. The recommendations aim to improve the trustworthiness of DoD's software supply and address areas in procurement, intelligence, quality and security assurance, acquisition, research and development, and the National agenda. The task force urges senior leaders in the U.S. Government to implement the recommendations in this report at the earliest opportunity.

Dr. Robert Lucky
Task Force Chairman

DEDICATION

Theoren P. (Trey) Smith, III

1954 – 2007

During the course of this Defense Science Board Task Force our colleague, Trey Smith, passed away from a long illness. Although Trey was not able to see this final report, his leadership and thoughts pervade it.

> *Trey Smith was Chief Technology Officer (CTO) of Science Applications International Corporation (SAIC). Before joining SAIC in 2002, Trey served in many executive roles at Cable & Wireless (C&W), Road Runner, Compaq Computer Corporation, and IBM. Prior to joining IBM, he conducted research at AT&T Bell Laboratories.*

> *Trey was a graduate of Brown University where he earned a Ph.D. in physics. He received a bachelor's in physics from the U.S. Naval Academy and served in the U.S. Navy as a nuclear engineer and division officer.*

> *Trey is survived by his wife Kathy, his daughters Kristen, Julia and Kimberly, and his son, Theoren.*

TABLE OF CONTENTS

Executive Summary v
Introduction 1
Findings 9
    Structure of the Software Industry 9
    Talent Base in Software 10
    Computer Science Education in the U.S. and Abroad 11
    The Growing Complexity of Cyberspace 13
    Software in the Department of Defense 16
    The Threat to DoD Software 21
    The State of Software Assurance in the DoD Today 30
    Existing DoD Information Assurance Policies to Address Software Vulnerability ... 35
    Ongoing Efforts in Software Assurance 36
    Supplier Trustworthiness Considerations 39
    Other National Security Acquisition Models 40
    Detecting Vulnerabilities 40
    Best Practices for Code Development 45
    Security Response Processes 47
    Software Assurance in Computer Science Education 48
    Conclusion 49

Recommendations 51
    DoD Procurement of COTS and Off-Shore Software 51
    Increase U.S. Insight into Capabilities and Intentions of Adversaries 51
    System Engineering and Architecture for Assurance 53
    Improve the Quality of DoD Software 55
    Improve Tools and Technologies for Assurance 56
    More Knowledgeable Acquisition of DoD Software 59
    A Research and Development Program on Assurance is Needed 65
    The National Agenda for Software Assurance 67

Appendix 69
    Terms of Reference 71
    Task Force Membership 73
    Descriptions of Example Software-Intensive Programs in DoD 75
    Task Force Findings and Recommendations 82
    Briefings Presented to the Task Force 85
    References 87
    Abbreviations & Acronyms 89

List of Figures

1. Market Share Changes 2
2. Risk vs. Consequence 6
3. Comparison of Degrees Awarded 12
4. Moore's law as demonstrated by Intel microprocessors. 13
5. Software size (MSLOC) increases over time 14
6. Total Web Sites Across All Domains 15
7. IAVM Taskings Issued 31

EXECUTIVE SUMMARY

Software has become the central ingredient of the information age, increasing productivity, facilitating the storage and transfer of information, and enabling functionality in almost every realm of human endeavor. However, as it improves the Department of Defense's (DoD) capability, it increases DoDs dependency. Each year the Department of Defense depends more on software for its administration and for the planning and execution of its missions. This growing dependency is a source of weakness exacerbated by the mounting size, complexity and interconnectedness of its software programs. It is only a matter of time before an adversary exploits this weakness at a critical moment in history.

The software industry has become increasingly and irrevocably global. Much of the code is now written outside the United States (U.S.), some in countries that may have interests inimical to those of the United States. The combination of DoDs profound and growing dependence upon software and the expanding opportunity for adversaries to introduce malicious code into this software has led to a growing risk to the Nation's defense.

A previous report of the Defense Science Board, "High Performance Microchip Supply", discussed a parallel evolution of the microchip industry and its potential impact on U.S. defense capabilities. The parallel is not exact because the microchip fabrication business requires increasingly large capital formation – a considerable barrier to entry by a lesser nation-state. Software development and production, by contrast, has a low investment threshold. It requires only talented people, who increasingly are found outside the United States.

The task force on microchip supply identified two areas of risk in the off-shoring of fabrication facilities – that the U.S. could be denied access to the supply of chips and that there could be malicious modifications in these chips. Because software is so easily reproduced, the former risk is small. The latter risk of "malware," however, is serious. It is this risk that is discussed at length in this report.

Software that the Defense Department acquires has been loosely categorized as:

- Commodity products – referred to as "commercial-off-the-shelf" (COTS) software;
- General software developed by or for the U.S. Government – referred to as "Government-off-the-shelf" (GOTS) software; and
- Custom software – generally created for unique defense applications.

The U.S. Government is obviously attracted by the first, COTS. It is produced for and sold in a highly competitive marketplace, and its development costs are amortized across a large base of consumers. Its functionality continually expands in response to competitive market demands. It is, in a word, a bargain, but it is also most likely to be produced offshore, and so presents the greater threat of malicious modification.

There are two distinct kinds of vulnerabilities in software. The first is the common "bug", an unintentional defect or weakness in the code that opens the door to opportunistic exploitation. The Department of Defense shares these vulnerabilities with all users. However, certain users are "high value targets", such as the financial sector and the Department of Defense. These high-value targets attract the "high-end" attackers. Moreover, the DoD also may be presumed to attract the most skilled and best financed attackers—a nation-state adversary or its proxy. These high-end attackers will not be content to exploit opportunistic vulnerabilities, which might be fixed and therefore unavailable at a critical juncture. Furthermore, they may seek to implant vulnerability for later exploitation. It is bad enough that this can be done remotely in the inter-networked world, but worse when the malefactors are in DoDs supply chain and are loyal to and working for an adversary nation-state—especially a nation-state that is producing the software that the U.S. Government needs. The problem is serious, indeed. Such exploitable vulnerabilities may lie undetected until it is too late.

Unlike previous critical defense technologies which gave the U.S. an edge in the past, such as stealth, the strategic defense initiative, or nuclear weaponry, the U.S. is protected neither by technological secrets nor a high barrier of economic cost. Moreover, the consequences to U.S. defense capabilities could be even more severe than realized. Because of the high degree of interconnectedness of defense systems, penetration of one application could compromise many others.

In a perfect world there would be some automated means for detecting malicious code. Unfortunately, no such capability exists, and the trend is moving inexorably further from it as software becomes ever more complex and adversaries more skilled. Even if malicious code were discovered in advance, attributing it to a specific actor and/or knowing the intent of the actor may be problematic. Malicious code can resemble ordinary coding mistakes and malicious intent may be plausibly denied. The inability to hold an individual accountable weakens deterrence mechanisms, such as the threat of criminal charges, or even separation of the individual or entity from the supply chain.

Task Force Conclusion

The Department of Defense faces a difficult quandary in its software purchases in applying intelligent risk management, trading off the attractive economics of COTS and of custom code written off-shore against the risks of encountering malware that could seriously jeopardize future defense missions. The current systems designs, assurance methodologies, acquisition procedures, and knowledge of adversarial capabilities and intentions are inadequate to the magnitude of the threat.

Task Force Findings

The Industry Situation

The software industry has become increasingly global as suppliers seek lower cost employees, access to a larger talent base, cultures conducive to highly-structured processes, and round-the-clock operation. The issue of foreign influence is only one of degree, because many companies develop code in multiple geographic locations and may embed code from other vendors, code from open source developers, or even code of unknown provenance.

While the United States still has preeminence in computer science, Asia is rapidly gaining. The United States retains a pool of talented computer scientists and engineers, but the natural tendency of the industry is to seek the lowest cost supply of talent. In recent years that has been primarily in India, while China and Russia are on the rise.

DoDs Dependence on Software

In the Department of Defense, the transformational effects of information technology (IT), joined with a culture of information sharing, called Net-Centricity, constitute a powerful force multiplier. DoD has become increasingly dependent for mission-critical functionality upon highly interconnected, globally sourced, information technology of dramatically varying quality, reliability and trustworthiness.

Software Vulnerabilities

The majority of software used in the Department of Defense is commercial-off-the-shelf product. Although the DoD takes advantage of the functionality and inexpensive pricing enabled by the huge market, this code has many weaknesses that are exploitable by even moderately capable hackers, who have been the beneficiaries of a culture that has produced an evolution of widely-disseminated and powerful tools for system intrusion.

The DoD does not fully know when or where intruders may have already gained access to existing computing and communications systems. The Moonlight Maze activities, which are classified and thus not detailed here, and numerous other data points demonstrate that the U.S. Government, and specifically the DoD computing systems, is a constant target of foreign exploitation.

The Threat of the Nation-State Adversary

In dealing with a nation-state adversary, the level of threat rises far above that posed by hackers. It can be assumed that the technological capability to craft actionable malicious code mirrors that of the United States own best computer scientists. Means and opportunity are present throughout the supply chain and life cycle of software development. While code developed in the United States is not immune from risk, the opportunity for an adversary is greatly enhanced by globalization.

A sophisticated adversary would have three possible aims in the exploitation of existing or planted software vulnerabilities – denial of service, stealing of information, and malicious modification of information. The outcome of any of these would also be accompanied by a loss of confidence in DoDs essential systems.

Awareness of the Software Assurance Threat and Risk

DoDs defensive posture remains inadequately informed of the sophisticated capabilities of nation-state adversaries to exploit globally sourced, ubiquitously interconnected, COTS hardware (HW) and software (SW) within DoD Critical Systems. Similarly, decision-makers are inadequately informed regarding the potential consequences of system subversion, and the value of mitigating that risk.

The Intelligence Community (IC) does not adequately collect and disseminate intelligence regarding the intents and capabilities of nation-state adversaries to attack and subvert DoD systems and networks through supply chain exploitations, or through other sophisticated techniques.

DoD does not consistently or adequately analyze and incorporate into its acquisition decisions what supply chain threat information is available.

Status of Software Assurance in the DoD

Software deployed across the DoD continues to contain numerous vulnerabilities and weak information security design characteristics. The DoD and its industry partners spend considerable resources on patch management, while gaining only limited improvement in defensive posture.

The evidence gathered during this study was insufficient to quantify the extent to which awareness and protection against the system assurance problem has permeated DoD systems and networks. The panel did however identify considerable variation in the extent to which the systems assurance problem is impacting next-generation DoD systems. That impact ranges from extensive with the introduction of inter-networked COTS and open source IT into of the Army's Future Combat System (FCS) program, to only slight in the United States Air Force (USAF) F-22 program.

The DoD defensive efforts, implemented largely through decentralized execution, are difficult to synchronize to achieve a coordinated enterprise effect. DoD has not effectively allocated assurance resources to address the systems assurance problem, nor has it designed its systems and networks to mitigate this problem in the face of the capabilities of nation-state adversaries.

The primary process relied upon by the DoD for evaluation of the assurance of commercial products today is the Common Criteria (CC) evaluation process. The task force believes that Common Criteria is presently inadequate to raise sufficiently the trustworthiness of software products for the DoD. This is particularly true at Evaluation Assurance Level-4 (EAL4) and below, where penetration testing is not performed. Nonetheless, Common Criteria evaluation is an international program, well established, and not easy to change.

Ongoing Efforts in Software Assurance

Software assurance is receiving attention at a number of federal agencies and laboratories, including the DoD, National Security Agency (NSA), National Institute of Standards and Technology (NIST), and Department of Homeland Security (DHS). Within the DoD a Software Assurance Tiger Team has been studying the problem and has developed a comprehensive strategy for managing risk through system engineering, source selection, design, production, and test. The key element of risk management in this strategy is the prioritization of criticality among system components and subcomponents, with special procedures and attention placed on the system components determined to be most critical to mission success.

Supplier Trustworthiness Considerations

It is not currently DoD policy to require any program, even those deemed critical by dint of a Mission Assurance Category I status, to conduct a counterintelligence review of its major suppliers, unless classified information is involved. Supplier trustworthiness enters into existing DoD acquisition processes primarily for protection of classified information and for research technology protection. From a systems assurance perspective, supplier trustworthiness should consider adversarial control and influence of the business or engineering processes of the supplier, as well as the ability of the business and engineering processes to prevent outside penetration.

Finding Malicious Code

The problem of detecting vulnerabilities is deeply complex, and there is no silver bullet on the horizon. Once malicious code has been implanted by a capable adversary, it is unlikely to be detected by subsequent testing. A number of software tools have been developed commercially to test code for vulnerabilities, and these tools have been improving rapidly in recent years. Current tools find about one-third of the bugs prior to deployment that are ever found subsequently, and the rate of false positives is about equal to that of true positives. However, it is the opinion of the task force that unless a major breakthrough occurs, it is unlikely that any tool in the foreseeable future will find more than half the suspect code. Moreover, it can be assumed that the adversary has the same tools; therefore, it is likely the malicious code would be constructed to pass undetected by these tools.

The task force believes that the academic curriculum in computer science does not stress adequately practices for quality and security, and that many programmers do not have a defensive mindset. While many vendors methodically check and test code, they are looking for unintentional defects, rather than malicious alterations.

Government Access to Source Code

It is tempting to consider having the U.S. Government take the source code of a commercial product and run its own vulnerability assessment tools against it. However, there are a number of legal, ethical, and economic barriers that make this an unattractive proposition, particularly from the point of view of the vendor. License agreements forbid reverse engineering of source code, vendors worry about the loss of intellectual property, and perhaps most importantly, they worry about the cost of supporting the actions and findings of a team of outsiders not familiar with the design and implementation of such hugely complex programs. Some of these worries are lessened when the testing is done by an independent laboratory.

Conclusion

All of the considerations just listed seem to point to an intractable problem. The Nation's defense is dependent upon software that is growing exponentially in size and complexity, and an increasing percentage of this software is being written off-shore in easy reach of potential adversaries. That software presents a tempting target for a nation-state adversary. Malicious code could be introduced inexpensively, would be almost impossible to detect, and could be used later to get access to defense systems in order to deny service, steal information, or to modify critical data. Even if the malware were to be discovered, attribution and intent would be difficult to prove, so the risk for the attacker would be small.

Against this backdrop of potential disaster, practical experience and belief paint a picture of aggravating and continuous software problems, but not ones that are lethal. However, there are some systems on which, to varying degrees, life depends (e.g., power, health). In this sense, DoD systems are among the most critical because their national security mission is often measured in fatalities, and failures that would be innocuous in another context can be lethal and lead to mission failure.

If the attacker cannot be deterred and its malware cannot be found, what is to be done to provide assurance that DoD software will perform in mission-critical situations? Although there never will be an absolute guarantee, software assurance is really not about absolute guarantees but rather intelligent risk management. The risk of vulnerable software can be managed through a suite of processes and mitigation strategies detailed in the Task Force recommendations, and this risk can be weighed against the attractive economics and enhanced capabilities of mass-produced, international software.

Task Force Recommendations

Acquisition of COTS and Foreign Software

DoD should continue to procure from, encourage and leverage the largest possible global competitive marketplace consistent with national security.

The DoD must intelligently manage economics and risk. For many applications the inexpensive functionality and ubiquitous compatibility of COTS software make it the right choice. In acquiring custom software the increased risk inherent in software written offshore may sometimes be worth the considerable cost savings. The task force recommends that critical system components be developed only by cleared U.S. citizens.

Increase U.S. Insight into Capabilities and Intentions of Adversaries

The intelligence community should be tasked to collect and disseminate intelligence regarding the intents and capabilities of adversaries, particularly nation-state adversaries, to attack and subvert DoD systems and networks through supply chain exploitations, or through other sophisticated techniques.

DoD should increase knowledge and awareness among its cyber-defense and acquisition communities of the capabilities and intent of nation-state adversaries.

Offensive Strategies Can Compliment Defensive Strategies

The U.S. Government should link cyber defensive and offensive operations to its broader national deterrence strategies, communications and operations, treating adversarial cyber operations that damage U.S. information systems and networks as events warranting a balanced, full-spectrum response.

System Engineering and Architecture for Assurance

DoD should allocate assurance resources among acquisition programs at the architecture level based upon mission impact of system failure. The task force endorses the strategy and methods to accomplish this as developed by the DoD Software Assurance Tiger Team and validated by the Committee on National Security Systems (CNSS) Global IT Working Group.

DoD cannot cost effectively achieve a uniformly high degree of assurance for all the functionality it uses across many and varied mission activities. Allocating criticality of function levies a requirement for assurance of that function, and also of those functions that defend it. Systems identified as critical must then allocate criticality at the sub-system and assembly level.

To properly allocate scarce assurance resources, DoD must allocate criticality at the system-of-systems and enterprise architecture level. This analysis should occur early within the life-cycle, and should render a prioritization decision no later than Acquisition Milestone A, to allow programs of record to appropriately respond to their criticality.

Improve the Quality of DoD Software

The DoD can effectively raise the "signal-to-noise ratio" against software attacks by raising the overall quality of the software it acquires. If there were fewer unintentional bugs in software, the visibility of deliberate malware would be increased. While general improvements in information assurance (IA) will not, per se, prevent a determined attacker from corrupting the software supply chain, there are several compelling benefits in improving the overall assurance/security-worthiness of COTS.

A sophisticated adversary would have to work harder to introduce an exploitable vulnerability instead, as is currently the case, of relying upon the plausible deniability of a common programming error to avoid attribution of malicious intent. Furthermore, a sophisticated adversary would have less confidence that its malware would remain undetected, invisible in a world containing far fewer distracting vulnerabilities. That uncertainty could be a deterrent in itself.

Improve Tools and Technology for Assurance

Improve Trusted Computing Group (TCG) Technologies

The Trusted Computing Group initiatives, centered on the Trusted Platform Module (TPM), provide a means for containing intrusions into separated information domains. Each chipset that implements the Trusted Platform Module embeds a unique identifier. Cryptologic verification of this identity is required when access to system assets is requested. TPM may help ensure that only approved and signed code is run, thus reducing the risk of unapproved code being installed.

NSA and others have identified a number of improvements and complementary practices that would strengthen TCG-compliant systems, including privacy-preserving attestation, virtualization, and architectures that provide richer software assurance measurement and monitoring capabilities.

Improve Effectiveness of Common Criteria

Currently, the official DoD-wide evaluation/validation scheme is the National Information Assurance Partnership (NIAP) based upon the Common Criteria. The reality today is that it would be far easier and more effective to improve Common Criteria than to invent a new scheme specific to the DoD or to DHS.

A number of ways to strengthen Common Criteria are discussed in the Recommendations section of this report. Among these suggestions are crediting vendors for the effective use of better development processes, including the use of automated vulnerability reduction tools and automated tools for vulnerability analysis—during evaluations at levels four (EAL4) and below. Validation schemes should also reduce artificial artifact creation and rely upon artifacts that are generated by the development process.

Improve Usefulness of Assurance Metrics

There is a natural tension between the U.S. Government's need to know the security-worthiness of what they procure and a vendor's need to avoid disclosing particular vulnerabilities. One way to satisfy both needs would be to develop a weighted index of the security-worthiness of software. A weighted score could be generated via testing based on some combination of the utility of the tools itself, the amount of code coverage of the tool, and the test results against a particular product. The entire development process should also be evaluated.

More Knowledgeable Acquisition of DoD Software

DoD should implement a scalable supplier assurance process to assure that critical suppliers are trustworthy. No product evaluation regime in effect today provides insight into a vendor's real development processes and their effectiveness at producing secure and trustworthy software – so the software assurance challenge for DoD is to define an evaluation regime that is capable of reviewing vendors' actual development processes and rendering a judgment about their ability to produce assured software.

The DoD acquisition process should require that products possess assurance matching the criticality of the function delivered. Furthermore, the DoD should require that all components should be supplied by suppliers of commensurate trustworthiness, and in particular, that all custom code written for systems deemed critical be developed by cleared U.S. citizens.

The collective buying power of the U.S. Government is such that it can force change on its suppliers to a degree no other market sector can reasonably do. The DoD, working in collaboration with the Office of Management and Budget (OMB), DHS, and other Federal agencies, can help to change the market dynamic through both positive and negative incentives so that they get better quality software, and to make better risk-based and "total cost"-based acquisitions.

Research and Development in Software Assurance

DoD should establish and fund a comprehensive Science and Technology Strategy and programs to advance the state-of-the-art in vulnerability detection and mitigation within software and hardware. The goals of the classified and unclassified research and development (R&D) investments in assurance should be to develop the technology to effectively take accidental vulnerabilities out of systems development and to improve Trusted Computing Group technologies in order to bound most risks of intentionally planted software. This program should monitor what markets are delivering, identify gaps between what the market is delivering and what DoD needs, and fill this gap.

INTRODUCTION

The 1990s saw a change from closed economies to free market mechanisms and a desire of more than half the world's population -- in India, China and the former Soviet Union – to compete economically. This has had a dramatic impact on the distribution of world income and caused dramatic changes to mature economies. It is clear that China, India, Russia and Brazil are no longer simply participating in low-end manufacturing, but are full-fledged economic participants in the global market place, with burgeoning internationally-competitive and internationally-employable populations. This transformation of the global marketplace is linked to the expansion of information technology into nearly every facet of life. Although reliance on IT is both beneficial and irreversible, it also creates significant opportunities for threat actors, in particular the nation-state adversary to control fundamental systems, networks, and critical infrastructure.

To place the issues of globalization in context, it is useful to consider both the current U.S. role in software development and world trends in this area. As of 2006, the United States held a clear leadership role in software development, as it did in hardware production not long ago. In 2003 the United States represented 65 percent of world-wide employment in the IT services sector, according to a McKinsey Global Institute study. Despite fears following the dot-com bust and rumors of massive outsourcing and off-shoring of software jobs, current U.S. employment in the various software disciplines is actually above the highs of the dot-com era. What then is the concern about globalization?

Much of the infrastructure required for globalization of software employment was created in the 1990s and earlier. This includes the astounding growth and spread of the Internet as a communications and distribution medium; the building and over-building of broadband fiber world-wide, virtually eliminating carrying costs as a factor; the wide-spread adoption of wireless, allowing societies to leap-frog infrastructure limitations; the broad adoption of a few standard applications such as Oracle and Systems, Applications and Products (SAP), allowing start-ups overseas to focus on fewer skills when offering support; significant improvements in higher education in technical skills, particularly in India and China; and the creation of an overseas job market for U.S.-trained foreign nationals, allowing them to return home to relevant jobs.

Thus, while the United States remains a dominant player in global software development, trends clearly point to erosion, reducing the United States to mere competitiveness. These trends may be irreversible. Software is and will increasingly be developed world-wide, chasing both lower costs and an expanding talent pool. Further, some of the strongest future players in this market space are adversaries of the United States. In some cases, the adversarial basis may be purely economic, while in others the United States must account for the possibility of confronting adversaries either directly or indirectly in military conflict. A generation ago, U.S. adversaries feared the risk of using hardware and software developed in the U.S., as it might be used as a weapon against them. Similarly, the United States must now confront -- and plan for -- the reality that adversaries may well be supplying the key hardware and software on which the U.S. bases its military and economic superiority.

Characterizing the extent of globalization and the challenge that it presents to the United States is complicated by the number of issues presented. The key point from the standpoint of this study is that U.S. multinationals are seeking -- and China, India and Russia are providing -- software services, up to and including design services. The extent to which the United States is off-shoring manufacturing operations to the detriment of U.S. labor market interests is not addressed in this report. These countries are entering into the market to produce software products, and they are supporting U.S. corporations in delivering products.

Figure 1, from the "Innovation Consortium," showing the erosion of U.S. market share, typifies the market situation. The next tier of data concerns the underlying capabilities that are supporting these market changes. There is much debate regarding the details of the data, but the clear trending of all indices (patents issued, R&D, PhDs granted, etc.) is that U.S. predominance in IT technology is rapidly eroding to mere competitiveness. The discussion below regarding the engineering graduate data is instructive. However, it is difficult to interpret competitiveness statistics for countries totaling five to six times the population of the United States.

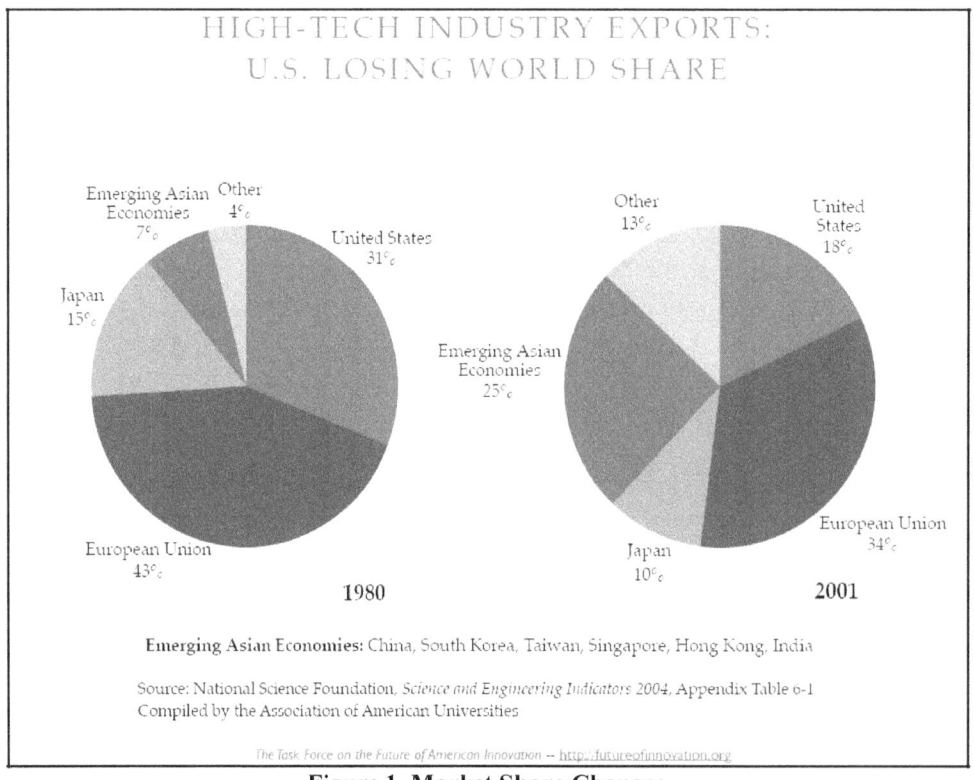

Figure 1. Market Share Changes

Any description of the state and trends in globalization of software must acknowledge the paucity of valid data on the subject, which makes any analysis tentative. Several factors contribute to the lack of valid data. Neither government nor industry has captured data for any period of time in this new arena. The data that exists has been captured inconsistently using differing parameters and definitions, and much of the data is captured with an agenda in mind, damaging its use for objective study. Most importantly, the whole issue of software globalization, in general, is so new that only a few years of data are available in any case. Thus, while there are clear trends discernable, it will be important to revalidate and refine these trends and the conclusions they drive before making radical changes in policy or practice. Data over the next few years promises to be more fruitful and exact. This contrasts markedly with the data available on hardware off-shoring, which is both easier to accumulate and has been available for some years now.

This Task Force considered the potential impacts of the action of numerous threat actors (hacker, criminal, terrorist, nation-states, and sophisticated nation-state adversaries, etc.) upon different elements of society (intellectual property, critical infrastructure, national security systems). However, it was the consensus of the group that ultimately the most difficult risk to address is nation-state adversary creation and exploitation of vulnerabilities that interfere with the DoD projection of military force. Further, DoD senior management remains inadequately informed about the reality of such a prospect, and even critical programs do not account for this risk.

The Impact of Globalization on Microelectronics

In 2003, the Under Secretary of Defense for Acquisition, Technology, and Logistics (USD(AT&L)) tasked the chairman of the Defense Science Board to stand up a task force to investigate the implications of the movement of semiconductor manufacturing and design capability offshore. The two principal concerns with this offshore migration were the DoDs inability to assure supplies of critical parts during times of crisis and the inability of DoD to assure design function based upon adversarial access to design and fabrication processes.

The recommendations of that task force were focused on ameliorating the threats posed by offshoring microelectronic fabrication, design, packaging, and testing. The initial focus of the study centered on issues and approaches for increasing confidence in U.S. access to critical semiconductor technology. However, as the study progressed, increased concern over the assurance of functionality of the technology itself rapidly emerged. This is best summarized by a quote from the report's introduction:

Introduction

> *"From a U.S. National Security view, the potential effects of this restructuring are so perverse and far reaching and have such opportunities for mischief that, had the United States not significantly contributed to this migration, it would have been considered a major triumph of an adversary nation's strategy to undermine U. S. military capabilities."*

Based upon the extent and potential impact that loss of the integrity/assurance of U.S. high performance microchip supply could have, the study additionally recommended chartering a comparable study for software. As stated in the study: "A strategy for achieving the above hardware counter-tamper objectives without a comparable strategy for software is of limited utility." Consequently, the task force recommended chartering a similar DSB study to investigate national security issues associated with rapid migration of software production, testing and maintenance overseas. This resulted in the formation of the Defense Science Board Task Force on Mission Impact of Foreign Influence on DoD Software.

Interrelationships between Software and Hardware Risks

The offense gets to pick the time, the place and the means to attack a target. The defense must be strong enough to withstand the strength of the attacker at the defender's weakest point. This is as true in information operations as it is in conventional warfare. The principal reason behind the microchip report's recommendation for the standup of this software study is embodied within this axiom of warfare.

From a defensive perspective, microelectronics and its associated software cannot be separated. While the offense may be able to attack either and meet operational objectives, the defense must be prepared for the offense to attack at the seam of software and hardware. If this offensive approach is done well and the defense examines the software and hardware only as independent elements, the offense is likely to go unnoticed until too late.

The microchip report provides the gamut of issues that affects both DoDs assured supply and the assurance of the components. The principal issues center on the fabrication facilities moving offshore due to extreme cost benefits and host country incentives. Once this market is centered in the U.S. opponents' back yards, other elements of the microelectronics process (e.g., design, test, and packaging) will follow. This provides significant access opportunities for the adversary and few, if any, new access alternatives for the U.S.

As conveyed in the present study, the issues surrounding the offshoring of software are diverse. Software design, development, testing, distribution, and maintenance can all be done more inexpensively offshore. As explained in Thomas Friedman's book, *The World is Flat*, global high bandwidth connectivity has allowed and encouraged third-world countries to compete effectively in the software arena where the cost of entry is low.

While some of the reasons driving microelectronics and software offshore are different, the impact on DoDs mission-critical systems, from incorporating foreign-developed parts, is the same. The U.S. is significantly enhancing the adversary's ability to access this technology in many of the life-cycle phases. This access provides opportunity to clandestinely modify the components that could negatively impact the functionality of mission-critical applications.

Risk: Threat, Vulnerability, and Consequence

The DoD now relies upon networked, highly-interconnected systems for many mission-critical capabilities, and this reliance is projected to increase. The software in these systems is the key ingredient that provides much of the increased capability delivered to the warfighter, just as it represents the key factor in increased productivity and new capabilities for industry today. For the DoD, this advanced technology is a force multiplier. Yet the reliability of that force multiplier is increasingly unknown due to the risks posed by a rapidly expanding supply chain that includes some of the Nation's most technologically sophisticated adversaries.

DoDs most highly sensitive systems are composed of defense-unique and highly-assured subsystems and components. However, other critical weapons systems, communication systems and support systems incorporate COTS or open source software (OSS). This software is likely to contain defects, as well as both accidentally and deliberately introduced vulnerabilities. Additionally, misconfiguration and/or poor maintenance of complex heterogeneous systems may also leave an organization vulnerable. All of these things are key components of system risk.

After vulnerabilities, threat is the second key component of system risk. Adversaries are well aware of the dramatic advantage that networked technology (Net-Centricity) brings to the U.S. military. They will assuredly seek to counter this advantage. To constitute a true threat, adversaries must have the intent, capability, and opportunity to attack DoD IT systems. If threat agents are able to identify and attack system vulnerabilities effectively, system failure will occur with the consequence that mission-critical capabilities may be compromised.

Global software development presents an opportunity for threat agents to attack the confidentiality, integrity, and availability of operating systems (OS), middleware, and applications that are essential to the operation of the U.S. Government and the DoD. The most direct threat is overt foreign corruption of software: insertion by the developer of malware, back doors and other intentional flaws that can be later exploited. A second threat is foreign adversaries' corruption of the commercial supply chain. Commercial development processes make no guarantees about the purity (or lack of corruption) of the supply chain, nor could they reasonably do so. The overall opaqueness of the software development supply chain and the complexity of software itself make corruption hard to detect. Furthermore, most companies are not actively looking for malicious introduction of suspect code, although there may be some collateral benefit from the fact that they do consider the provenance of code to ensure no infringement of intellectual property rights.

Figure 2 displays the relationship among threats, vulnerabilities, risks and consequences. Risk, in this case, is defined as the probability of system failure. Threat and vulnerability contribute to this risk. The consequence of system failure must be assessed based on mission impact and the criticality of the mission.

Figure 2. Risk vs. Consequence

Assessing systems according to the framework in Figure 1 encourages prioritization of resources toward IT systems where risk is high and the consequence of system failure is severe. Since risk and consequence are independent factors, decision makers can consider whether resources are most effectively employed in reducing risk (e.g., by insisting on government-supplied software in a weapon control system) or in mitigating the consequence through a redundant system (e.g., by fielding a medium range, armed Unmanned Aerial Vehicle (UAV) system in addition to an artillery system).

This framework illustrates that DoD can raise the level of IT system assurance by countering threats, by reducing vulnerabilities, reducing opportunities for vulnerabilities to be introduced (e.g., using cleared people) or mitigating the impact of vulnerabilities through redundancy, isolation and other techniques. For example, the threat component of risk can be countered by reducing the opportunities of threat agents to insert malicious code at points in the supply chain. System vulnerability can be reduced by detecting and correcting defects and malicious code. Vulnerable COTS or open system software modules can be defended by placing them within robust, well-defended architectures.

As this report points out elsewhere, cost and capability advantages will increasingly drive the DoD to incorporate into its systems COTS software and IT hardware components produced on foreign soil. The provenance of commercial IT may provide adversaries an easier opportunity to compromise DoD systems. Other links in a supply chain that includes production, installation and upgrades may provide additional opportunities for foreign influence. Such opportunities must be appropriately recognized and countered. Similarly, other elements of threat must be identified and countered. In addition to recognizing and reducing the opportunity element of threat, DoD and the Intelligence Community can neutralize adversaries through judicious countermeasures. Discovering and then eliminating or defending vulnerabilities addresses a major component of risk. At the same time these actions serve to increase the capabilities an adversary must possess to mount an effective attack.

The DoD and the broader national security community have recognized this problem and have identified some possible solutions. The DoD Software Assurance Tiger Team has developed a strategy and concept of operations (CONOPS), described later in this report, to manage these risks. The Committee on National Security Systems (CNSS) established a Global IT Working Group, which independently validated the work of the Tiger Team and other studies regarding the risks of globalization. The working group then developed a strategy applicable to the broader Federal community.

FINDINGS

Structure of the Software Industry

FINDING: The software industry, as well as the software talent base, is becoming increasingly global, and this trend appears to be irreversible.

The software industry comprises a diverse collection of organizations that cannot be simply categorized. There are large and often well-known software providers with widely distributed development facilities and smaller software providers whose development facilities may be completely located in one geographical area. There are system integrators who serve a critical function in integrating hardware and software and may write custom software as part of that process. Other companies (like financial institutions and military weapons providers) are not traditionally thought of as software houses, but in fact create large quantities of software as a routine part of their business. The abundance of software is reflective of its transformative value, while the diversity in the industry is reflective of the low barriers to entry. Indeed, even individuals with limited formal education in computer science can create commercial or not-for-profit software companies and cheaply distribute software around the globe.

The Task Force focused on the larger software companies, system integrators, and military providers who, as part of their normal course of business, provide the DoD with significant amounts of software. In many cases, these companies have foreign nationals working in the United States as well as geographically distributed software development processes. Moreover, these companies are increasingly off-shoring software development.

Market realities are such that few companies develop, ship, and support their software using only U.S. citizens located in the U.S. or, for that matter, only foreign developers located overseas. "Foreign influence over software" may thus be a matter of degree. The market reality is that many commercial software companies develop products in multiple geographic locales. Even a U.S.-based company of any size can and likely does have a worldwide presence if not worldwide development. In fact, venture capital firms routinely pressure software start-up firms to use overseas talent. Many companies may also embed code from other vendors (open source or code licensed from a third party) which itself may be of unknown or unproven provenance.

There are clear benefits to global development for the software industry that both directly and indirectly benefit DoD:

- First, offshoring gives companies access to a larger pool of talent (so that they can continue to produce world-class software). Presumably, having a healthy domestically-owned or domestically-controlled software industry, even if development is done globally, is preferable to having no domestic industry at all.

- Second, developing software overseas potentially results in much lower development costs, so companies can potentially supply more functionality - including functionality of interest to DoD – without raising prices. For example, given that DoD tends to have massive systems, the ability to build more scalable and reliable systems by investing the cost savings generated by offshore development is a tangible benefit to DoD.

- Third, there may be advantages to offshoring that provide a tangible benefit in terms of quality or security-worthiness. Better talent produces better quality software. Moreover, there is anecdotal evidence from the Year 2000 (Y2K) experience that developers in some countries may as a group be more willing to follow a structured development process (such as the Capability Maturity Model (CMM)), instead of arguing that development processes emphasizing security also "stifle creativity." Ultimately, security-worthiness is at least partly about quality, and quality is in part a result of talented people following strong development processes and paying attention to secure coding practice.

Talent Base in Software

The globalization of the software industry is driven not just by cost. Although off-shoring can reduce costs, because pay scales in other countries may be lower than those in the United States, it is likely that this pay gap will diminish over the long term. This is because one result of globalization is to increase opportunity and improve the standard of living in more countries, thus causing pay scales to increase in other economies as demand for quality workers increases. Indeed, over the last few years, the gap in pay for software developers has grown smaller, even if still substantial. Thus in the longer term it is not just about cost but access to talent, an issue ironically exacerbated in part by the government's own policies that limit the number of Visas issued to foreign IT workers.

An adequate supply of software talent is the key enabler for developing a sound technology base in software research and development. In order to evaluate the adequacy of the U.S. software industry's ability to meet DoDs needs for custom software, the Office of the Under Secretary of Defense for Acquisition, Technology and Logistics (OUSD(AT&L)) sponsored a Software Industrial Base Study (SIBS). The two-phase study was launched to assess the demand for software within DoD and the industrial base's ability to satisfy that demand.

Phase I of the study, completed in October 2006, found that the overall pool of software programmers appears to be adequate. However, there is a larger demand for the upper echelons of the software developer and management cadres, exacerbated by the fact that people trained in these specialty areas generally cannot easily transfer their skills across domains of expertise. In order to develop the specialized software that is most critical to the DoD mission, knowledge of the application domain is harder to obtain and more valuable to DoD than knowledge of a particular programming language or even knowledge of software engineering itself.

Software engineers who have expertise in defense-oriented applications are likely to be in greater demand in the DoD sector than in the commercial marketplace. Likewise, employers in the DoD sector are highly motivated to retain experienced software engineers because of the expense of training new ones in the relevant applications, as well as the cost and delay of obtaining a security clearance for a person entering from the commercial market. Phase I of the SIBS specifically highlights a key choke point in the pool of top-tier program management/software architect talent. Industry is trying to grow this cadre with training and career development strategies for high-potential employees, highlighting the critical need to coordinate government human resources strategies with industry. Phase II of the study is underway.

Computer Science Education in the U.S. and Abroad

The United States remains one of the world's most competitive arenas for computer science education, and the heart of the current U.S. workforce (ages 45-64) is still the best educated workforce compared to other countries. The United States is producing a significant number of engineers, computer scientists and information technology specialists, and remains competitive – though no longer dominant -- in global markets. The challenge for the United States over the next decade will be to retain its role as a global pacesetter in the education of engineering and scientific talent and thereby to sustain its legacy as a preeminent technological innovator. Programs in software engineering and other efforts to train thousands of new programmers with the needed analysis, design, development, and project management skills are not widely offered in the university and post-graduate education systems and, additionally, demand for these courses is dropping. Indeed, industry uniformly notes that educational institutions are not producing enough job candidates and those that they are producing lack the necessary grounding in security to develop products that address today's threats.

Varying, inconsistent reporting of problematic engineering graduation data has been used to fuel fears that the United States is losing its technological edge. In addition to the lack of nuanced analysis around the type of graduates and quality of degrees being awarded, these articles also tend not to ground the numbers of engineers in the larger demographics of each country. While the growth of offshoring activities in the IT sector, particularly in India and China, is real and increasing, the United States is not in the desperate state that is routinely portrayed.

Many of the studies on rates of graduating engineering students world-wide consider only the raw number of graduates, without factoring in either the accuracy of what is counted or the quality of the education. Such studies also evaluate only total numbers of graduates from each country, without considering per-capita

Findings

graduating rates per country. A recent report developed by graduate students of Duke University's Master of Engineering Management Program does attempt to consider the varying education standards of India and China compared to the United States, and does examine the data in the context of each country's total population.

The Duke study has determined that the massive numbers of Indian and Chinese engineering graduates reported in various studies include not only four-year degrees, but also three-year training programs and diploma holders, as shown by Figure 3 below. This is particularly significant for the outsourcing debate, as the competition off-shore for talent focuses on only the best educated and experienced individuals among four-year graduates, and those holding advanced degrees.

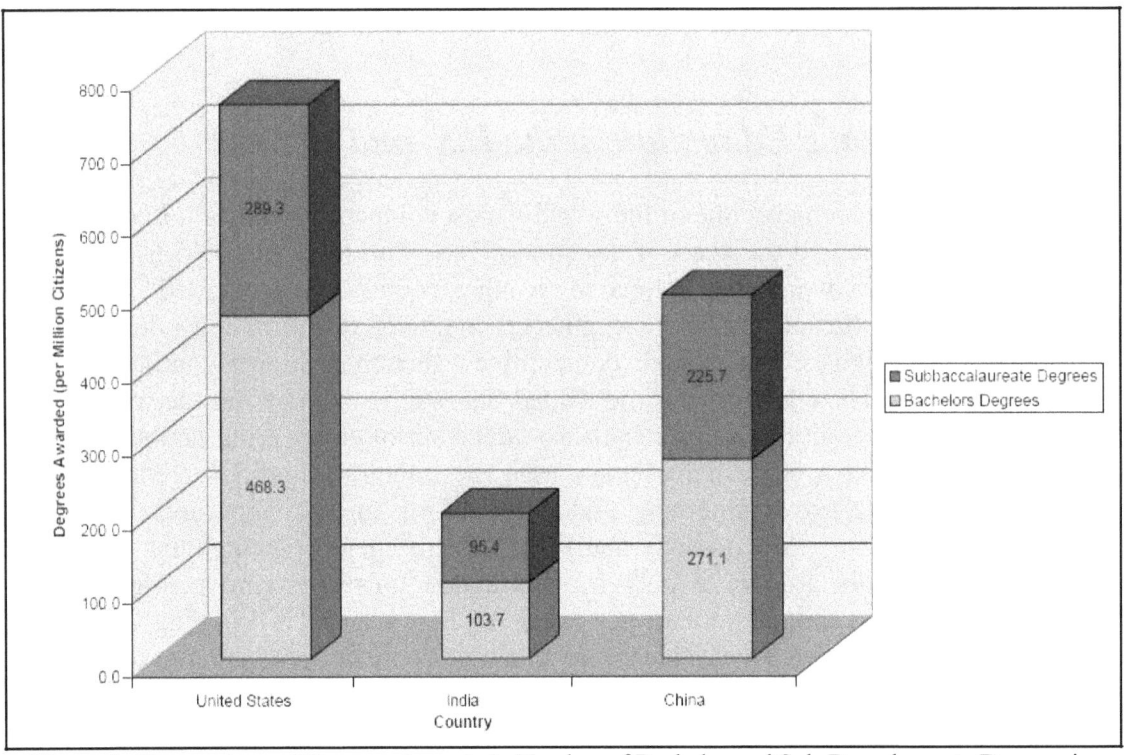

Figure 3. Comparison of Degrees Awarded. Number of Bachelor and Sub-Baccalaureate Degrees in Engineering, Computer Science, and IT Awarded per Million Citizens in 2004.

The Duke Study notes that per capita the United States produces 750 technology specialists for every million citizens, compared with 500 in China and 200 in India. A comparison of like-to-like data suggests that the United States produces a significant number of engineers, computer scientists and information technology specialists, and remains competitive in global markets. The great majority of engineers involved in outsourced professions hold a minimum of a four-year degree. Given the higher degree requirements for outsourced professionals, one could argue that approximately half of China's and India's annual engineering and IT graduates are capable of competing in the global outsourcing environment.

However, a recent McKinsey global labor market study argues that this estimate is far too generous. McKinsey concluded that only 10 percent of Chinese engineers and 25 percent of Indian engineers can compete in the global outsourcing market, after factoring in language skills, educational quality, cultural fit, job accessibility, as well as the attractiveness of domestic non-outsourced jobs.

The Growing Complexity of Cyberspace

There is an increasing complexity in micro-electronics, software, and computer networks as shown by the following data.

Micro-Electronics

Micro-electronic devices have been exponentially increasing in complexity over time. Intel co-founder Gordon Moore noted this in 1965 and stated what became known as Moore's law -- that the number of transistors on a chip about doubles in a short fixed time period. His original formulation predicted a time period of one year for the years 1965-1975[1], and later stated the period of two years[2] (though 18 months has been widely quoted). Simply looking at one microprocessor line – Intel's – shows that this prediction of exponential growth in micro-electronics' number of transistors per chip has held true all the way through today (Figure 4 below).[3]

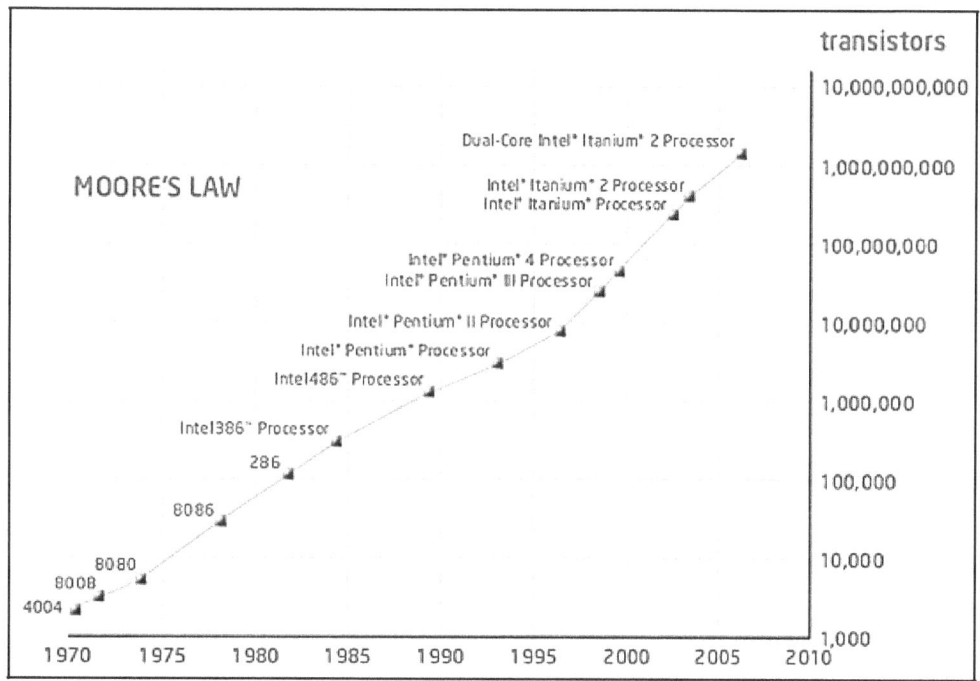

Figure 4. Moore's law as demonstrated by Intel microprocessors.

---

[1] Moore 1965
[2] Intel 2007
[3] Intel 2007

Findings

Exponential growth in the number of components has been shown in many other microelectronic products such as Random-Access Memory (RAM) chips (with corresponding sizes[4]), and it is widely expected that this trend will continue for some time to come.

The rules of physics cannot be violated, so no exponential growth can occur forever. Even Moore has stated that doubling every two years "can't continue forever."[5] Regardless, microelectronic components are expected to continue to grow in complexity for many years to come.

Software

Software tends to grow significantly in size over time, illustrated in Figure 5. Software source lines of code (SLOC) values publicly available for products such as Microsoft Windows and Red Hat Linux give strong evidence that this trend is continuing.[6] This increasing size brings with it increasing complexity.

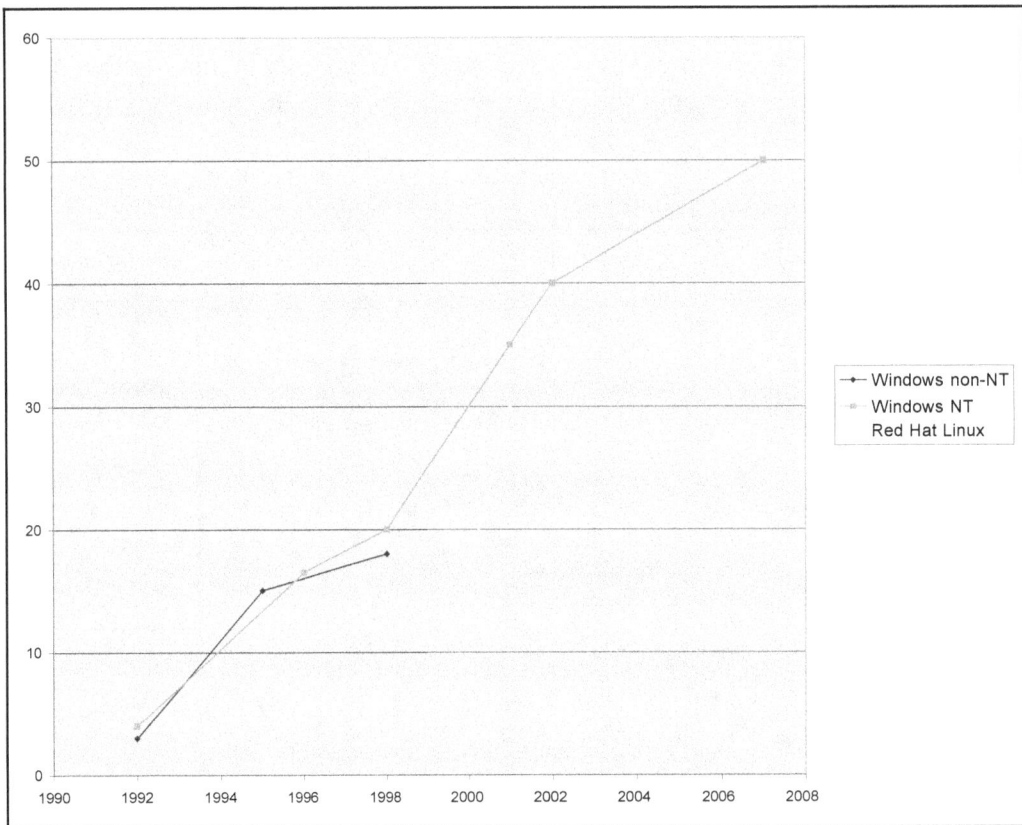

Figure 5. Software size (MSLOC) increases over time.

---

[4] Dix 2004
[5] Dubash 2005
[6] Schneider 2000, McGraw, Associated Press 2007, Wheeler 2000, Wheeler 2001

Networks

The number of interconnected computers has grown rapidly as well.

Netcraft has tracked the total number of web sites on the (public) world wide web since 1995. Originally it only tracked the number of unique names with a website. As the number of "inactive" sites grew (e.g., sites which existed but did not yet post content), it began separately tracking the number of active sites. Figure 6 shows a steady growth in the number of publicly-accessible web sites from 1995 on.[7]

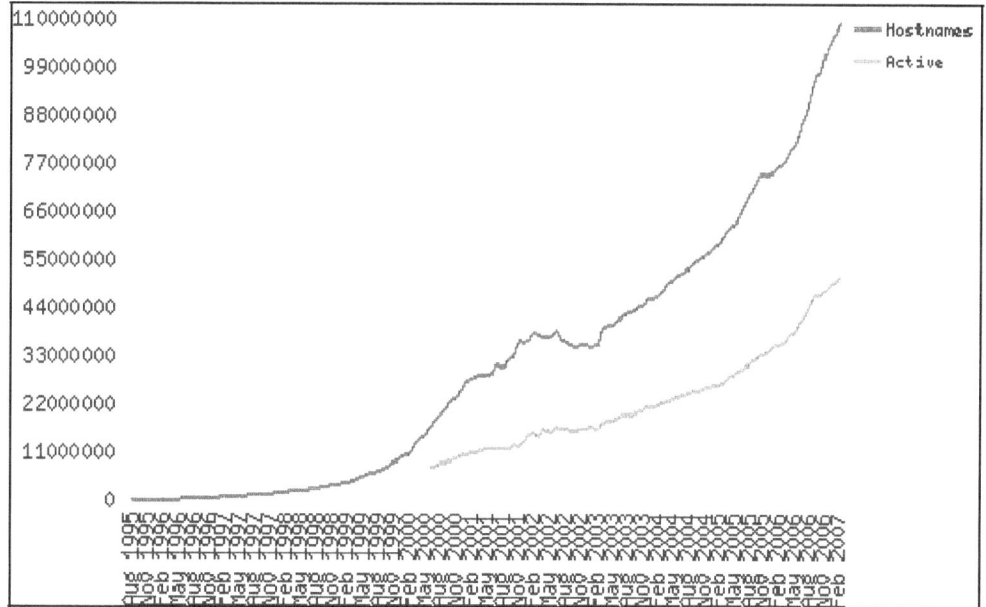

Figure 6. Total Web Sites Across All Domains August 1995 - February 2007.

Other studies have attempted to determine the number of hosts on the Internet (whether or not they are web servers), including Lottor's Request for Comments (RFC) 1296, Zakon's[8], and the Internet Systems Consortium (ISC). These studies confirm that the number of computers interconnected through the Internet has been rapidly rising for decades. Computers and systems in DoD, and throughout the U.S. Government and U.S. critical infrastructure, have been swept along in this trend, creating avenues of attack across DoD systems and networks.

---

[7] Netcraft 2007
[8] Zakon 2006

Findings

Software in the Department of Defense

FINDING: DoD has become increasingly dependent for mission-critical functionality upon highly interconnected, globally-sourced, information technology of dramatically varying quality, reliability and trustworthiness.

Over the past several decades, global competition in the information technology industry has dramatically reduced the cost of hardware and software IT products. At the same time, advances in networking and communications technologies have made connectivity through commercial protocols ubiquitous. In the DoD, the transformational effects of such technologies have created a Net-Centric environment, where information and functionality are shared and interconnected to a high degree. While Net-Centricity is a powerful force multiplier, the increasing dependency upon this capability creates and amplifies the risk of vulnerability to unintentional or maliciously-induced system malfunction.

Size and Scope of DoD Software

To bring context to this discussion, it is useful to consider the sheer size and scope of DoDs dependence on software, and the increasing pace of that dependence. Software has been growing in the dimensions of size, complexity, and interconnectedness, each of which exacerbates the difficulties of assurance. This and the following sections of this report discuss these trends.

The amount of software contained within DoD systems has been growing exponentially.[9] It is estimated that the DoD cleared industrial base, including approximately 70,000 programmers, software engineers and computer engineers, meets the annual demand for software written for DoD. There is no definitive source of data on the amount of code required annually by DoD. However, the Center for Strategic and International Studies (CSIS) study did develop an estimate of 60 million software lines of code (SLOC) (new and maintenance code) required annually by the national security community in 2006.

The amount of software included within DoD systems is considerably larger than this estimate when COTS software integrated into DoD systems and networks is included. The COTS content of DoD Systems and networks varies considerably. Command and Control (C2) and Mission Support systems are probably predominantly COTS. The Future Combat System (FCS) represents a significant intermediate point: Of the total estimated 63.8 million SLOC for FCS, 12.3 million will be written for the program, 14.6 million SLOC will be reused and 36.9 million SLOC will be COTS. On the F22 Program, which leverages high assurance COTS developed under the multi-independent levels of security program (MILS) for certain mission-critical elements of the platform, software is predominantly custom, totaling approximately 4 million SLOC.

---

[9] CSIS, 2006

DoDs increased risk comes not only from the continuing shift to software developed overseas, but from the sheer volume and dependence DoD has on all types of software to meet mission requirements. Some measurements relating to software are the size and scope of DoD IT activities, as indicated by the following FY2006 data points (from the Department of Defense Chief Information Officer (DoD CIO).

- The DoD FY06 IT Budget is $30 billion.
- DoD has some 5 million Computers.
- DoD has issued approximately $10 million Common Access Cards (CAC).
- DoD Operates in some 600,000 facilities in 6,000 locations in 146 Countries.
- DoD has three global Intranets (NIPR, SIPR, and Joint Worldwide Intelligence Communication System (JWICS)).

Foreign Dependencies in DoD Software

Although there are no firm figures, it appears that custom DoD software is developed predominantly by U.S. citizens, many of whom hold clearances. The impact of globalization on the pedigree of source code within DoD systems is driven by the globalization of the supply chain for COTS, called "packaged software," to distinguish such software from software services and from embedded software within commercial hardware components. There does not appear to be direct information regarding the percentage of this software that is globally sourced. Recent studies indicate that global sourcing (from high wage to low wage countries) of SW Engineering talent within the SW Product sector can be expected to increase from 7 percent (41,000 full time equivalents FTE)) to 18 percent (116,000 FTE) over the period 2003 to 2008.[10] Thus, ignoring direct imports of foreign origin COTS, COTS purchased from U.S. companies can be expected to become increasingly foreign sourced. At a minimum, the lines of code would approach the percentage of global sourced labor (18% in 2008). Because, by function, R&D is among the most easily globally sourced, this figure can be expected to increase over time barring supply shortages. Moreover, this number does not include COTS products developed by foreign manufacturers and sold to DoD or incorporated into DoD systems and networks by the DoD industrial base.

Varieties of Software used in the DoD

Software is developed within many different business models. Broadly, software may be divided into Custom, GOTS, and COTS. In fact, the issue is more complex than this, and in each of these cases there are closely related types (e.g., freeware and open source software), and in each case software developed within one business model may be incorporated into another. For example, many COTS developers and DoD custom code developers may also embed open source or code from other developers, which itself may be of unknown or unproven provenance. Because custom software can be well characterized and controlled, most security

---

[10] McKinsey, 2006

and reliability risks for the DoD are frequently associated with open source and COTS software. However, this is not always the case, as, for example, when rapidly prototyped custom software finds its way into the operational setting and soon becomes indispensable. At this point in time almost all IT environments are composed of "mixed code."

Custom Software

Custom software is specifically designed for an individual customer. While it may later be cleared or released for broader use, in general an individual customer procures custom software for a specific purpose, intends to be the sole user of this software, and takes concrete steps to protect the proprietary nature of the software and its derivatives. Examples might include quantitative algorithms for evaluating hedge funds for a financial institution or avionics software for an advanced aircraft manufacturer.

GOTS refers to custom software developed for the government for one purpose that is later reused "off the shelf" by contractors and other government agencies and integrated into new government systems and applications. GOTS software may be developed by the technical staff of a government agency or by an external entity with funding, oversight and specification from the agency, or by a combination of both. The DoD has established and employs a large and stable industrial base for both custom and GOTS software. Where specified or required, contractors and vendors develop or support 100% of their software in the United States using only U.S. citizens who are vetted, and shipping code that has been reviewed and tested to meet specific security, reliability and operational requirements.

Commercial-Off-The-Shelf (COTS) Software

COTS software is used "as-is." COTS products are designed to be easily installed and configured to interoperate with existing system components. Almost all software bought by the average computer user and much of the software used by the U.S. Government and the DoD is COTS. Examples include operating systems, database management systems, email servers, application servers, and office product suites. Because it is mass-produced, one of the major advantages of COTS software is the relatively low cost of acquiring, maintaining and achieving technology refresh. Given these low costs and the competitive forces at work, COTS software producers may or may not know, manage or track the provenance of their software, except to the extent needed to ensure that the necessary license are obtained for embedded components. In addition, they generally do not make source code available, so supplier identity and software content is often blurred by the reuse of legacy code, subcontracting, outsourcing, and use of Open Source Software (OSS).

Open Source

In general, open source refers to software whose source code is freely available for use, modification and redistribution by the public, often without royalties. It is usually developed through public (open) collaboration with the result that there may be little or no knowledge of the motivations or loyalties of many of its developers. Examples of OSS include the Linux kernel, the Apache web server and the Mozilla Firefox web browser.

Most major systems integrators use OSS in their systems. In addition, several open source organizations have emerged (e.g. The Open Source Initiative) that provide certification and several companies deliver open source solutions as their core business. Freeware and shareware preceded open source and were available for use, modification or distribution at no cost for an unlimited or limited time, respectively. As with some open source now, freeware and shareware, and derivatives, could also be distributed at no cost. The result has been that freeware and shareware are embedded within countless software programs, modules and systems.

Managed Services

Another area of increased opportunity and risk for DoD is the rapid increase in the use of managed services. The switch to this kind of support is projected to increase rapidly nation-wide over the next decade, and DoD will not remain immune from the trend. The greater risk inherent in managed services stems from the need to give unparalleled access to the service provider of services as part of the contractual arrangement. DoD has begun to address this concern in policy, requiring that DoD elements considering a contract for managed network services get a counterintelligence assessment of the proposed provider in order to identify risks from contracting with companies controlled by countries or organizations of concern.

The Growing Complexity and Interconnectedness of DoD Software

Software complexity is growing rapidly and offers increasing challenges to those who must understand it, so it comes as no surprise that software occasionally behaves in unexpected, sometimes undesirable, ways. Sometimes the seeds of this behavior are indigenous to a specific software module but sometimes the cause of the anomalous behavior is "environmental"—*i.e.*, it stems from some other software co-habiting the immediate process space, or the larger application space, or on other machines in the network space.

In addition to what are traditionally considered applications, there are numerous support processes or daemon processes running in the background. This leads to "unconscious" interdependencies that may cross process boundaries. The most common example of such interdependency is the buffer overflow, in which a program fails to check input sufficiently with the result that program code embedded in the input gains control of the process. If there is a critical function to perform, it must be ensured that it does what, and only what, was intended. However, these interdependencies allow what may be a less critical function to disrupt the operation of a more critical function, unless great pains are taken to keep them separated.

Findings

A buffer overflow may not occur in a critical application, but in a co-resident application. More importantly, the operating system may not safeguard against other processes that are intended—by their designer, but perhaps not by the user—to interact. Like most security functions, they work reasonably well to keep the honest people from doing something wrong. Was that buffer overflow in a neighboring program mere inadvertence, or malice aforethought?

DoD decision-makers are intentionally trading off confidence in the assurance of the critical application for the cost, performance, and availability benefits of COTS. In these systems, the functionality in excess of that required to meet the requirements of the critical application is staggering. This excess functionality creates unnecessary risk, pushes the envelope of DoDs evaluation ability, provides ample access opportunities for opponents, and creates domain isolation "by-pass" paths. Users making decisions to include excess functionality do so with an understanding of the benefits without a corresponding understanding of the associated risks.

Note that economics and technology exacerbate these problems of interdependency and interconnectedness. General purpose platforms running a multitude of applications seem like a "best buy," and programmer productivity rises dramatically by re-use of legacy code -- where provenance is neither known or of concern to the developer.

The very complexity of this problem ensures no simple solution, but this cannot serve as an excuse to ignore it. Attending carefully only to a critical application is no grounds for confidence on the overall performance of the system. The entire environment must be of concern.

Examples of Software Development in the DoD

The Aegis System introduced in 1980 required fewer than 2 million lines of code. The F-22 aircraft introduced in 2005 requires perhaps double that, while the F-35 aircraft scheduled for introduction around 2010 will have an estimated 12 to 14 million lines, and the Future Combat System planned for around 2015 will require an estimated 16 to 18 million lines of custom code, and some 60 million lines of COTS, GOTS and open source. Correspondingly, in the F-4 introduced in 1960, software controlled some 8 percent of functionality by software, while software will control 80 percent of the F-22.

The Task Force received briefings from three warfighting programs: FBCB2-Blue Force Tracking, F-22, and FCS. All three programs are viewed as important to the DoD war fighting mission. These three programs are described in Appendix C.

Broadly speaking, the Task Force found that the three programs varied widely regarding the extent of their incorporation of COTS, their reliance upon general purpose Internet protocols (IP), and in their attention to risks posed by their supply chains. FCS and F-22 had processes to consider supply chain risks within the lifecycle. That said, none of the programs have in place a process to analyze the supply chain for the kinds of risk posed by a nation-state adversary.

The Threat to DoD Software

Prior to the development and broad adoption of the Internet, adversaries were in many cases forced to gain close access to their IT targets. This generally meant that in order to be operationally effective, the opponent needed to have the resources and accesses typically associated with a nation-state intelligence service. These activities were typically limited in scope, not easily scalable, and not a core operational technique. Additionally, during this period, the motivation for applying operational resources within this area was limited, as the value of the intelligence residing inside these systems was usually not high grade ore (with some notable exceptions). With the introduction of the Internet, this rapidly changed. Not only did this bring into the game the hacker level adversary, it also significantly increased the utility and scalability of targeting IT. The volume of the gold and the grade of the ore simultaneously increased.

Net-Centricity offers the DoD immense opportunity to increase effectiveness of force projection, but also is an attractive target for those attacking the confidentiality, integrity, and availability of DoD data and systems. Future warfare environments will involve information sharing by coalition forces via the Secret Internet Protocol Router Network (SIPRNET) as well as the Non-classified Internet Protocol Router Network (NIPRNET), thereby expanding greatly the number of persons with access and the potential for mischief.

A combination of factors has produced a high-risk condition: system designers placing insufficient emphasis on security; an increase in the number and sophistication of attackers; the growing abundance and value of information on the Internet; and a significant increase in utilization and dependence on the technology. Hacking techniques are widely disseminated and automated, while "defensible coding" is not. And, as the adage goes, developers must find all bugs, hackers need find only one. Thus, the advantage rests with the attackers. The defense has not begun to respond from an institutional perspective. From a defensive perspective, the preponderance of thought is that the challenge/threat is modest and DoDs defenses are adequate. After all, there has never really been a "smoking gun" that has caused concern or could not be explained away. This perspective undermines a core component of U.S. national security.

The DoD does not fully know when or where intruders may have already gained access to existing computing and communications systems. The Moonlight Maze activities, which are classified and thus not detailed here, and numerous other data points demonstrate that the U.S. Government, specifically the DoD computing systems, is a constant target of foreign exploitation. Further, the Eligible Receiver series of exercises have demonstrated an ability to penetrate and exploit selected DoD systems. And recently, the Commerce Department revealed that attacks by

Chinese hackers forced one of its bureaus to cut off Internet access and discard virus-infected computers.

FINDING: Adversaries' use of low-level cyber attack techniques to exploit weak information assurance controls and vulnerable software has led to successful attacks upon sensitive but unclassified (SBU) systems and networks of DoD and the defense and commercial industrial base. This has occurred notwithstanding extensive DoD efforts in security and information assurance.

Attacks on computer systems are designed to breach (1) confidentiality, (2) integrity, or (3) availability.

Malevolent modification of the system or its components is the most challenging to detect and protect. It is a domain that likely interests the United States most sophisticated potential adversaries far more than most other Internet aggressors, and also a domain in which the most sophisticated potential adversaries may have a decade of experience from tampering other systems (fielded and in development). Net-Centric systems surely will attract sophisticated adversaries who can subvert the supply chain to replace or alter software or hardware, recruiting well-placed insiders and exploiting single-string dependencies. Even if the provenance of each part of an IT component can be determined, with precision, the complexity of the technology makes the challenge of discovering malicious constructs within those component(s) remarkably difficult. In fact, a classified experiment conducted in the mid-1980s demonstrated the overwhelming challenges of discovering subversion structures within microprocessor/software based systems of the time.

Knowing how much to trust the software to do what, and only what, it is supposed to do is far from a science. Worse, having once lost confidence in a system, it is unlikely that such confidence will be readily restored. Military contests are, in fact, contests of will, and if an adversary can cause the U.S. Military to doubt its own capabilities, the United States determination will be lessened and their coercive powers enhanced. A military legend is instructive: A combatant commander is said to have remarked that he would not go into battle unless his blood supply was assured and, in this context, he meant not only assured physically, but that the information about the blood supply was equally assured—i.e., trustworthy. The reality is that DoD does control its blood supply, but its control of the information system surrounding the blood supply could be placed in doubt.

There are those with a deep understanding of the range of offensive options that can be arrayed to target IT and the corresponding impact of the associated security compromise. Unfortunately, for those in the information assurance camp, the problem seems overwhelming and while they quickly accept the magnitude of the problem and turn to seeking solutions and strategy, little effective progress has been made on the defensive side. Consequently, there is a large and growing gap between what a sophisticated offense can achieve and what a good defense can protect. By way of example, the offense has been hard at work for at least twenty years (e.g., "The Cuckoo's Egg"), but security efforts did not massively increase until 9/11.

Recent computer/network intrusions into DoD systems have resulted in increased national interest in this problem. Numerous studies have been conducted and more are in progress, partly to ascertain the impact of such attacks. Congressional interest is growing. The Weapons of Mass Destruction (WMD) commission recommended starting a new Presidential commission focused on this national security challenge. Ironically, had it not been for an adversary's use of simple and detectable offensive approaches, the DoD may not have had the current level of attention on this most important problem.

Opportunistic and Targeted Vulnerabilities

Although the DoD needs to plan for a wide range of threats, attacks on DoD IT systems fall into three groups: opportunistic -- launch of a general attack on the Internet which, by extension, impacts DoD operations; specific targeting of DoD systems by someone without sponsorship or political motive; and specific targeting conducted by a dedicated (and potentially well-funded) adversary. The differences are important, especially when there is widespread reliance on commercial software.

Opportunistic attacks involve actors who are not specifically targeting the DoD, but whose actions could nonetheless affect military operations. Worms (which may cause denial of service attacks) and spam (which can significantly impact costs and productivity) represent this type of threat. Significantly, improvements in commercial software development practices – along with proper management of IT systems by implementing appropriate risk management and defense-in-depth strategies -- can eliminate this threat or greatly reduce its impact.

Targeted attacks may come from different sources and be prompted by different motives. Throughout the 1990s, for example, hackers would often attack government systems because they made interesting targets and provided bragging rights. In some of those cases, the actor neither had a broader political motive nor state sponsorship. In these cases, better security – although perhaps providing more of a "challenge" -- may serve to drive the attackers elsewhere (much like a good home security system, the goal is not to defeat a concerted effort but to make the target less attractive). More importantly, better security may include better auditing and evidence collection, thus ultimately resulting in the prosecution of the hacker and the creation of a meaningful deterrent effect.

The deterrent value of computer security is relatively small today, in large part because the number of hackers actually identified and prosecuted successfully is small, and criminal penalties are sometimes too lenient to provide real deterrence. In some countries--both friend and foe--laws prevent or significantly reduce prosecutions. But commercial needs are driving better authentication on the network, governments are increasingly collaborating on cyber-crime investigations (note the recent U.S. ratification of the Council of Europe Cyber-Crime

Convention, the first major cyber-crime treaty) and government-industry partnerships around Internet safety and cyber-crime prevention and response are growing. Thus, there is hope that greater deterrence can be created through better investigative practices and international coordination.

In contrast to targeted attacks by non-sponsored individuals, sponsored attacks raise a different set of problems. Assuming an adversary is sufficiently motivated to devote meaningful resources to a structured effort, better DoD IT security will not serve as an adequate protection. Additionally, there is no meaningful penal response, since a sponsoring country will not aid the victim country in investigating and prosecuting the actors. Simply put, organized attackers using tradecraft can probe in perpetuity, looking for zero-day bugs (i.e., bugs unknown even to the product vendor), unpatched vulnerabilities, or misconfigurations of systems. No COTS development process will deliver a product capable of withstanding a dedicated nation-state attack -- attackers can reverse engineer patches faster than system administrators can test them and apply them, and configuration of complex heterogeneous systems remains difficult, even as vendors increasingly automate the configuration process for their own products. Moreover, vendors do not prohibit users from clicking on attachments, although more protective mechanisms (e.g., sandboxing) are now being deployed. State sponsored or organized transnational efforts can result in protracted attacks where information is slowly compromised.

Relative Risks in COTS and Custom Software

COTS software development environments can be more easily penetrated than custom development environments because:

- COTS software contains other packages (not developed in-house) that are not tracked or otherwise flagged as "not invented here." Consider the case in which a developer downloads an encryption routine off the Internet that may not be implemented well or that has been deliberately corrupted, and nobody knows that this package was downloaded from someplace else. Software not known to be there in the first place will not be factored into a risk assessment. Obviously, this problem varies considerably from vendor to vendor, with significantly less risk coming from vendors with robust security development processes.

- The software is known to contain "not invented here" code, but it is of unknown provenance and unknown security-worthiness. For example, consider a case in which object code is shipped with a commercial product and the vendor does not have source. Some code may have been there so long nobody remembers where it came from. Without source, it may be difficult to vet security-worthiness.

Thus, while COTS development environments are more porous to attack than those of DoD custom development environments, subversion or tampering with components destined for the latter are more likely to achieve adversarial objectives.

In short, attacks on COTS are easier, but attacks on custom products are more likely to meet the requirements of the adversary.

The risk of damage from maliciously introduced vulnerabilities increases with the ease of adversarial access to the development environment. U.S. businesses and foreign businesses with development based in the United States are not immune to adversary manipulation by U.S. citizens or "outsiders" infiltrated into a domestic operation. That said, an adversary with "home court advantage" finds it safer and cheaper to manipulate the production and distribution of software products in its own control.

Reliability engineers have long recognized that frequent modification of software through patching can reduce reliability by introducing new flaws, even while some patching brings upgrades to fix reliability issues. The less obvious but more serious concern is the use of patches to deliberately introduce new exploitable vulnerabilities or to sustain old ones. As with any supply chain, the entirety of the supply chain is only as secure as its weakest link and ongoing maintenance is, unhappily, a fact of life. Seldom is the entire supply chain under a single auspice, so the continuing maintenance permits new actors to enter in a variety of guises.

The Nation-State Adversary

System administrators and information assurance officers are often pre-occupied with everyday "hacker and patch/update" activities. Many see these as the only threats, overlooking the far greater risks posed by the approaches/techniques of the high-end threat posed by sophisticated adversaries.

Even if a target is either not connected or has no known exploitable vulnerability, the adversary must find a way to get close to the target or to introduce vulnerability through a life-cycle operation. It is this vector that separates the hacker from the sophisticated adversary. It is this vector that provides abundant operational opportunity for the sophisticated adversary who can take advantage of the global IT market from an access perspective.

It is the consensus of the Task Force that sophisticated nation-state adversaries will examine the possibilities of exploiting supply chain weaknesses in DoD acquisition of critical systems. These adversaries will then seek to exploit such weaknesses to degrade or deny capability of critical systems if they determine it is their best option to ensure damage while remaining hidden. These adversaries will attack to their best advantage, and their advantage improves significantly when software development moves under their control through the globalization of software development. The Task Force sees exploitation of software and hardware development or production as a key area for such adversaries. Bottom line: globalization of software development, where some of the United States adversaries are writing the code the DoD will depend upon in war, creates a rich opportunity to damage or destroy elements of the war fighter's capability.

FINDING: The enormous functionality and complexity of IT makes it easy to exploit and hard to defend, resulting in a target that can be expected to be exploited by sophisticated nation-state adversaries.

Scope and Character of the Threat from a Nation-State Adversary

Unlike the terrorist, the nation-state adversary seeks to create subversions that persist even years into the future, while remaining hidden, as a key objective of a nation-state adversary is to put in place reliable offensive capabilities well in advance of need. The ability to subvert a system through supply chain exploit, while remaining hidden, is greatly enhanced by deep knowledge of the context within which the tainted subsystem will operate (e.g., system design, intended use) or what components it will include.

The nature of espionage is changing in the age of globalization, but in many cases, U.S. characterizations of an intelligence organization are mired in the past. This becomes a problem for those who are a target of such an organization and still have defenses in place to deal with the old threat. During the Cold War, clandestine technical operations were primarily used to support human recruitment, but today these technical activities are becoming more of the main act. In fact, human operations are increasingly being used to support technical operations. For one embracing the old paradigm, this is somewhat counter-intuitive.

This transition is largely due to rapidly growing utilization of technology to create, store, manipulate, and communicate secrets. Secrets are no longer the exclusive domain of safes and the minds of targets. Additionally, based upon DoDs overwhelming dependence on technology, the target is not always the secret but increasingly the target is the integrity and availability of the technology itself to support its intended mission. As mission-critical systems become progressively more reliant upon technology, a denial of service attack becomes extremely attractive. At the core of most of this advanced technology is software.

FINDING: Nation-state adversaries are able to employ a full spectrum of offensive intelligence trade-craft, including attacks that subvert the supply chain, to damage or defeat mission-critical systems. The risk from such supply chain exploits can only increase as larger portions of the supply chain become more accessible to the adversary through global sourcing.

A common, but incorrect, perception is that the only way for a sophisticated adversary to attack a network-accessible target is with cyber tools. While having a negative impact on the confidentiality, integrity, or availability of the cyber target is the operational objective, the adversary has a rich array of tools at its disposal: surreptitious entry, spies, signals intelligence (SIGINT), clandestine technical collection, cyber, foreign partners and the use of cover companies. This array of tools allows the sophisticated adversary to conduct a targeting assessment to determine how it can meet its operational requirements while minimizing costs and risks.

Once this assessment is completed and a high-level plan developed, the full spectrum of modern day espionage is engaged: operational access options are examined; partners are approached; special technology is developed; and where appropriate, legal/policy constraints are examined. This opponent gets to pick the time, the place, and the combination of methods conducted within a veil of secrecy to achieve their objectives. The synergistic and mutually supportive nature of these tools/approaches can yield powerful offensive results. This offensive paradigm is often ignored, dismissed as not real, or deemed too difficult to handle. For mission-critical applications, this may well lead to mission failure.

As the importance of life-cycle access to a sophisticated adversary's success becomes more evident to the DoD, a reasonable conclusion to draw would be to significantly limit these access opportunities to the adversary by mandating all software and hardware be designed, developed, tested, distributed, and maintained only by trusted agents. Unfortunately, even if these agents were U.S. companies, the door for access opportunities would be far from closed to the adversary. By design, a sophisticated foreign intelligence service has the ability to operate clandestinely within the United States backyard.

There are other factors that compound the magnitude of the defensive challenge. First, it is well understood that the COTS strategy for DoD is an important part of its business model for mission success. A corollary to this strategy is that provenance of the piece parts of the software is increasingly foreign. As discussed above, each of the foreign pieces provides an access opportunity for the adversary.

Second, even if one knew with precision the provenance of each part, the complexity of the technology makes the challenge of discovering malicious constructs within the component(s) remarkably difficult.

Third, the technical and operational ease of altering software coupled with the scalability and inexpensive character of the approach make it a very desirable offensive strategy.

A summary that characterizes the gap between the location of a sophisticated offensive operation and the United States best defensive, implemented strategy is found in The Digital Dimension:

> *The defensive challenges are daunting! Today there is no clear approach to effectively offsetting the advantages of the offensive adversary. What are the intrinsic characteristics and environmental factors of the defensive situation that have resulted in a growing gap between offense and defense? They include technology complexity; impracticality of comprehensive evaluation; design, production, and maintenance functions performed globally; imbalance of focus and competing aspects among confidentiality, integrity, and availability; insufficient*

insight into the offensive investments, approaches, and organization of opponents; lack of coordination and cooperation between offensive and defensive elements; no national-level ownership; and the absence of a national research - investment strategy.[11]

The Risk in Low-Level Attacks

FINDING: Nation-state adversaries' penetration of sensitive-but-unclassified systems and networks could allow them to steal system information or tamper with system artifacts, enabling them to target existing and future DoD Systems.

The software that is being developed for military purposes, as well as the software that can be predictably expected to be used by DoD or other high value targets, is stored as data within networks of commercial industry and the DoD industrial base. When the networks of these organizations are penetrated, such software can be stolen. This is a significant problem for intellectual property protection and acquisition program protection. However, frequently confidentiality attacks are not as pernicious as integrity and availability attacks. Software and other elements (e.g., Application-Specific Integrated Circuit (ASIC) designs) can be tampered with (back doors, logic bombs), or other defects inserted. This can expose DoD systems yet to be fielded to operational risk.

Knowledge of the Nation-State Adversary Threat

FINDING: DoDs defensive strategies and techniques remain inadequately informed of the sophisticated capabilities of nation-state adversaries to exploit globally sourced, ubiquitously interconnected, COTS HW/SW within DoD Critical Systems, and the potential consequences of system subversion.

Currently, there are no effective means for incorporating critical information regarding major cyber attack/subversion into DoD mission planning. Military exercises do not include consideration of system failure due to cyber attack or other covert means of subversion as an element of the exercise. There is no strategic consideration of command and control failure within mission planning as has been required historically for critical command and control (e.g., within nuclear command and control). Yet the high levels of interconnectedness among IT applications in C2 and sensitive-but-unclassified networks yield ever increasing opportunities for large-scale failure. These failings are a natural consequence of the historical disconnect between those who attack systems and those tasked with defending them.

As described above, there is a tendency to take a narrow view of the nation-state adversary threat, believing that it is limited to cyber operations only. This arises from a number of factors that prevent those acquiring, defending and operating

---

[11] Sims 2005

systems and networks from getting full information regarding the scope of the threat.

1. Real knowledge of what the adversary is capable of is tightly held within the offensive operational community for fear that disclosing such information will hinder the ability to use those capabilities.

2. There is no clear understanding within the offensive operational community of the defensive actor's need to know such information, and the risk to national security if the defense does not understand the adversaries' capabilities.

3. The amount of offensive information to be disclosed and the number of defensive players with a need to know requires that highly compartmented information be so "watered down" during declassification, that it may be too vague to be actionable.

FINDING: The Intelligence Community (IC) does not deliver timely, actionable intelligence regarding the intents and capabilities of nation-state adversaries to attack and subvert DoD systems and networks through supply chain exploitations, or through other sophisticated techniques.

In 2002, the Intelligence Community Acquisition Risk Center (ICARC) led an effort to review and improve collection requirements for supply chain threat in both the signals intelligence (SIGINT) and human intelligence (HUMINT) arenas. It appeared to have little or no impact or follow through. There is little evidence that such proposed changes were adopted, and even less evidence that these requirements, if adopted, would have pushed far enough up the resources chain to have impact. What is needed is a thorough revamping of the collection requirements to capture the capabilities and clarify the intent of nation-state adversaries to exploit the supply chain. Also required is some education for the collectors and taskers to ensure that there is an understanding of the criticality of such information. For instance, information is needed on the movement of intelligence officers from and to commercial companies, as well as ties that the intelligence services develop with commercial entities, whether through infiltration, investment or partnering.

Without a better understanding of the capabilities and intents of nation-state adversaries, DoD and the U.S. Government run the risk of investing heavily in areas of protection of IT software and hardware when in fact there is no plan or capability to exploit it. Alternatively, without a clear understanding of the threat, program managers are likely to refuse to expend funds on protecting their systems from an unproven threat, thereby enabling a sophisticated adversary to do maximum damage at little risk or cost.

Findings

FINDING: The risk of undetected subversion of custom software is considerably greater than the corresponding risk for COTS.

Adversaries' Model of Risk Management

This report details and will later recommend a greater utilization of risk management techniques. While it is understood that there will never be a perfect defense, it is not well understood how to perform effective defensive risk management in today's IT environment. At a minimum, one must characterize systems, keep these characterizations current, know the impact of system/component compromise, understand the intentions, motivations, and capabilities of all adversaries, discover system vulnerabilities and threats, apply/develop techniques to close the vulnerabilities, and exercise incident response plans in preparation for an event.

Often overlooked is the importance of the defense's understanding of the risk management factors and decision factors of the opponent. A defensive posture is enhanced by addressing these factors. The offense must ask:

- What is the utility/intelligence gain of conducting the operation?
- What is the cost (dollars, people, opportunity, etc.) of conducting the operation?
- What is the likelihood of the opponent's defense detecting my operation?
- What is the probability of the opponent attributing this operation to our organization/country?
- What are the consequences of the operation being discovered and attributed to us?
- What are alternate operational approaches to meeting our requirements?

By more deeply understanding the adversary's answers to these questions and the weight that they attach to each factor, the U.S. will be better positioned to optimally apply both active and passive defensive resources. As long as the impact of defensive compromise remains high, these assessments, investments, operations, and strategies must be developed and acted upon. This must not be a static exercise. It must be continuous and relentless.

The State of Software Assurance in the DoD Today

In the security context, software assurance is best defined as the level of confidence that software is free of vulnerabilities either intentionally or unintentionally designed or inserted during the software development and/or the entire software lifecycle. The true goal of software assurance is to prevent the exploitation of software, thus increasing its dependability and defensibility.

Vulnerabilities in DoD Code

FINDING: Software deployed across DoD continues to contain numerous vulnerabilities and weak information security design characteristics. DoD and its industry partners spend considerable resources on patch management, while gaining only limited improvement in defensive posture.

The DoD Information Assurance Vulnerability (IAV) Management (IAVM) process was created to prepare and rapidly disseminate mitigating actions for potentially critical software vulnerabilities to DoD Components. IAVM notices have three criticality levels:

- IAV Alert (IAVA) – most critical – a vulnerability posing an immediate and potentially severe threat to DoD systems.
- IAV Bulletin (IAVB) – less critical than IAVA, but pose a threat to DoD systems.
- IAV Technical Advisory (IAVT) – less critical than IAVB.

Figure 7, on the next page, shows that the total number of vulnerabilities resulting in IAVM notices has increased each year, as has the number of the most critical IAVAs and IAVBs. The number of IAVTs decreased in 2006, but since the total number of notices increased, reported vulnerabilities tended to be more critical in 2005 and 2006:

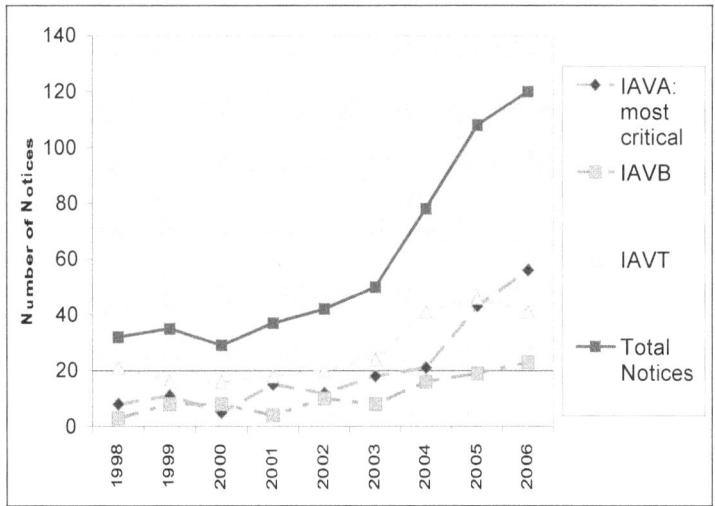

Figure 7. IAVM Taskings Issued

These IAVM figures only include the most critical vulnerabilities and do not count multiple vulnerabilities addressed within a single alert; the true number of publicly-known software vulnerabilities is far greater. Over 17,000 vulnerabilities have been publicly published on the Internet from 1998 through mid-2006; over 6,000 of them could cause root-level compromise, yet only 423 IAVM notices have been published in that period. Thus IAVM notices do not include 93 percent of the publicly-known root-level compromise vulnerabilities, and do not include 98 percent of the publicly-known vulnerabilities.

There are simply more vulnerabilities than DoD can handle: Issuing an IAVM for each release of a new remote root compromise vulnerability would render the IAVM program ineffective and cost DoD precious resources. Alerts provide priority of effort in vulnerability mitigation to limit burden and resources imposed.[12] In addition, for mission-critical or complex software, programs are unlikely to be patched quickly regardless of the severity of the issue, precisely because the systems are so critical or complex. Such conditions mandate significant testing before patch deployment. Other reports have also found that the number of software vulnerabilities reported annually is rapidly increasing.[13]

Common Criteria

FINDING: The primary processes relied upon by DoD for evaluation of the assurance of commercial products, Federal Information Processing Standards (FIPS) and NIAP Common Criteria Evaluation Process, do not address software vulnerabilities except for higher assurance levels (if at all). Moreover, these processes do not scale to the volume of software critical to the DoD mission.

In 1997, NSA and NIST established the National Information Assurance Partnership (NIAP) with the goal to increase the level of trust consumers have in their information systems and networks through the use of cost-effective security testing, evaluation, and validation programs. Using the Common Criteria as its standard[14] for evaluating information assurance products, the NIAP Common Criteria Evaluation and Validation Scheme was established in 2000. The Common Criteria enables the validation of security functionality, and improves confidence that the product is free of vulnerabilities through an increasing set of evaluation assurance levels (EAL1 through EAL7). At levels above EAL4, nation-specific methodologies are required; however, at EAL4 and below, the evaluations/validations are recognized by countries through the Common Criteria Recognition Arrangement. NIAPs prime mission is to evaluate/validate information assurance products, not specifically software, but a substantial amount of software is evaluated at varying functionality and assurance levels through NIAP and the Common Criteria Recognition Agreement for use within the DoD.

Common Criteria (CC) evaluates the extent to which a product's security features comply with a formal description of those features. The description may be either a (government) user-provided "Protection Profile" or a developer-provided "Security Target." Evaluations against a Protection Profile assess the extent to which a product complies with user requirements – evaluations against a Security Target merely evaluate the extent to which the product does what its developers say it does.

---

[12] Lawrence 2006
[13] ISS 2007, Messmer 2007,. Lemos 2006
[14] ISO 15408

Because most product vulnerabilities arise at the implementation (code) level, and because evaluations at EAL4 and below focus only on the product's design documentation, Common Criteria evaluations at EAL4 and below have no real bearing on confidence that a product is reasonably free from vulnerabilities.

The NIAP was formally assessed through a Government Accountability Office (GAO) study and informally through a variety of draft DoD documents. One advantage of using the Common Criteria is that it replaced multiple, country specific evaluation criteria that forced the vendors to evaluate the same product, at great expense, in multiple venues. With the addition of the Common Criteria Recognition Arrangement, vendors received definite cost benefits, recognition for their evaluated information assurance products within a more global market place, and the amount of labs evaluating products expanded while adding competition for their evaluation services. On a case-by-case basis, other advantages have included improvements in product developmental practices, and some reduction in vulnerabilities especially during the design of security features. Criticisms of Common Criteria-based schemes are that they are expensive, require artifacts that are not produced until well after product design and implementation, do not substantially reduce implementation-level vulnerabilities when using today's software development practices, and lack thorough penetration analysis at EAL4 and below. There is, therefore, an effort to promote a new regime that addresses these concerns.

Capability Maturity Model

Late in 1986, the DoD asked the Software Engineering Institute (SEI) (a DoD sponsored federally funded research and development center focused on improving the state of the practice of software engineering) to develop a method for assessing the capability of software contractors. The Software Engineering Institute developed a process maturity framework that would assist organizations in improving their software engineering processes. In September of 1987 the SEI published the process maturity framework and a maturity questionnaire to be used as a simple tool to identify areas where an engineering organization's software process needed improvement. This initial work was based on the premise that the quality of a software product is largely determined by the quality of the process used to develop and maintain it.

Over the next several years the DoD, defense contractors, and the SEI gained experience in using the maturity framework and questionnaire, and interest in software process and software process improvement grew substantially. The SEI used its experience in conducting software process assessments and extensive feedback from government and government contractors to mature the framework and to evolve it into the Capability Maturity Model (CMM) for Software.[15] The CMM provides a structured approach to apply the principles of Total Quality Management to software development and maintenance. It helps software

---

[15] Paulk 1993

organizations gain control of their software processes, identifies key practices that organizations should adopt to achieve improvement, and provides a framework for process improvement that allows organizations to develop higher quality software at lower cost. Studies of the effectiveness of the CMM conducted by the SEI[16] and others[17] clearly demonstrate that organizations can gain substantial benefits from investing in software process improvement: earlier detection of software defects during the development process, reduction of defects in deployed software, reduction in calendar time to develop software systems, and overall productivity improvement.

Since 1993, momentum behind the process improvement concepts has increased dramatically and the concepts embodied in the CMM have been applied to other aspects of the engineering process: e.g., managing the engineering workforce using the People Capability Maturity Model[18], improving the software acquisition process using the Software Acquisition - Capability Maturity Model (SA-CMM)[19], improving systems engineering process using the Systems Engineering Capability Maturity Model[20], and improving organizational and project management using the Integrated Product Development Capability Maturity Model.[21] To respond to the growing interest in maturity models and the need to integrate them in a coherent way, in 1998 the Office of the Under Secretary of Defense for Acquisition and Technology, with co-sponsorship from the National Defense Industrial Association, initiated the Capability Maturity Model Integration (CMMI) project to establish an overall framework that could accommodate current and future models. This work has resulted in the release of the initial CMMI product suite in 2002 and most recently in the CMMI for Development Version 1.2.[22]

While use of the CMMI products by DoD contractors and others continues to grow substantially and organizations that adopt the models continue to show substantial improvements in their ability to develop higher quality products with improved schedules and lower costs[23], it is important to note two key points for the purposes of this report.

First, while the CMMI product suite includes rigorous methods to assess an organization's capabilities, and many individuals are trained and authorized to conduct quality assessments, the assessments produced are assessments of organizations' potential to create high quality products, but not a guarantee that they will do so on any particular project. Process assessments by themselves do not examine the outputs of any development effort and are therefore silent with respect to the quality attributes of any particular product. A positive process assessment

---

[16] Herbsleb 1994
[17] Dion 1993, Lipke 1992, Wohlwend 1993
[18] Curtis 2003
[19] Cooper 2002
[20] EIA 1998
[21] DoD 1996
[22] CMMI Product Team 2006
[23] Goldenson 2003

finding lowers the risk that an organization will produce a low quality product but the quality of the product itself must be assessed using other methods.

Second, while today's models help organizations improve the overall quality of the products they produce, they provide no guidance for dealing with specific security concerns such as protecting against the intentional insertion of malicious code during the development process. Further research and development is needed to augment today's models with specific security improvement practices that would be effective at protecting against the threat of errors that would inadvertently introduce security vulnerabilities or attacks that would lead to intentionally embedded malicious code. The current CMMI models are a good foundation for these new practices and building the new practices on that foundation would help promote faster transition of the practices to the large and growing community of DoD contractors and other software developers that have already invested in adopting the CMMI models. Also, as mentioned above, process assessments must be augmented with improved product assessments to provide the level of assurance needed for critical systems.

Existing DoD Information Assurance Policies to Address Software Vulnerability

Several national security and DoD information assurance policies provide guidance applicable to assuring software for the DoDs most critical systems. DoD Directive 8500.1 on information assurance specifies that information assurance requirements be included during the acquisition of all DoD information systems. This policy also directs compliance with the National Security Telecommunications and Information Systems Security Policy #11, which requires that all information assurance and information assurance-enabled IT hardware, firmware, and software be properly evaluated and validated by specifically defined criteria. Additionally, DoD 8500.1 directs the thorough assessment of risks for software with limited warranties or software that is freeware or shareware.

Risk assessments are made within the certification and accreditation processes. The DoD is now using the DoD IA Certification and Accreditation (C&A) Process (DIACAP) for systems. DIACAP requires identification of the security requirements and an independent analysis certifying the completion of those requirements. Accreditation occurs when the residual risks are managed and the system is allowed to go operational. The definition of "system" varies. For the most part, a system is comprised of a number of hardware and software components. Information technology products should not be confused with the system itself, and the verification of the security requirements is resolved by analyzing the entire system and not merely the individual products of which it is composed. The role of software assurance within the overall certification and accreditation of systems is critical. Many of the security requirements will be implemented in software, or use software in conjunction with hardware support at the system level.

Although it is not practical to ensure that all software is vulnerability free, gaining confidence in the software development practices of the developers is crucial. A third party product evaluation process should provide usable information to make certification and accreditation both more effective and more efficient. Unfortunately, the current use of Common Criteria falls well short of this goal. In addition to product certifications, particularly critical systems may need to be formally modeled, formally specified and developed, thereby adding a mathematical basis to the development of the software. Assurance is additionally gained by independence in the certification and accreditation processes. The search for vulnerabilities should start with ensuring the proper security requirements, searching for design vulnerabilities, building vulnerability searches during the development of the code, and by having approved configuration management and change management procedures in place throughout the entire lifecycle. The level of assurance is also affected by the supply chain: Who built the code? Where was the software developed? What control processes were in place? How was it distributed? Who integrates it into the final product? Who administers the code? How are upgrades performed, and by whom?

Appendix C describes three large DoD programs that incorporate extensive software, and how issues of assurance were managed in these programs.

FINDING: The Task Force identified considerable variation in the extent to which the Systems Assurance Problem is impacting next-generation DoD systems. That impact ranges from extensive with the introduction of inter-networked COTS and open source IT into the Army's Future Combat System (FCS) program, to only slight in the USAF F-22 program.

Ongoing Efforts in Software Assurance

DoD programs are addressing the implications of the Systems Assurance Problem on a program-by-program basis. Currently, there is no comprehensive policy in effect to address this problem. Acquisition programs addressing the Systems Assurance Problem, such as FCS, while managing risk in a reasonable way, are assuming too much without adequate analysis. This is being encouraged by adoption of strategies or assumptions whose very applicability is highly doubtful (e.g., that the profit motive of COTS provider will weed out foreign intelligence front companies or that DoD can make blind purchases where vendors do not know their products are destined for DoD use). Managing the systems assurance problem is best accomplished by augmenting and amplifying activities not currently second nature within acquisition programs: enterprise architecture, systems engineering, information assurance, program protection, vulnerability detection, etc.

The DSB explored the security risks due to globalization in a DSB Task Force on "Globalization and Security," December 1999. DoD and the intelligence community have been steadily working the software and systems assurance problem. In 2002, the Intelligence Community Acquisition Risk Center (CARC) acquisition process of the CIA was broadened to include the entire IC, pursuant to

DCID 7/6. The ASD(NII)/DoD CIO established the Software Assurance Initiative in October of 2003. Based upon the findings and recommendations of the ASD(NII)/CIO Software Assurance Initiative, the USD(AT&L) and ASD(NII)/DoD CIO established the Software Assurance Tiger Team (described further in the next section). In addition, beginning in 2005, the Committee on National Security Systems (CNSS) established the Globalization of IT Working Group (GITWG) to examine the problem from a national perspective and develop a national strategy.

FINDING: DoD Critical Systems and networks remain vulnerable to the sophisticated capabilities of Nation-state adversaries. USD(AT&L) and ASD(NII)/DoD CIO, through the efforts of the DoD Software Assurance Tiger Team and the CNSS Globalization of IT Working Group (GITWG), have developed a comprehensive strategy and CONOPS for addressing the systems assurance problem within DoD and an approach for the Federal Government at large.

DoD efforts, culminating in the DoD Software Assurance Tiger Team, have led to a Systems Assurance Concept of Operations (CONOPS). Within the DoD System Assurance CONOPS, operational needs are translated into a requirement for assurance through a prioritization process, which identifies the critical systems. Acquisition programs for these critical systems employ systems engineering processes with measures to identify critical subcomponents, and then deploy designs to protect them. These critical subcomponents are then subject to more rigorous oversight, source selection, design, production, and test (e.g., F-22). Most importantly, such components are protected from components of lesser criticality by functions implemented with the same level of assurance as the functions they are defending.

To compete for system procurements, industry suppliers must meet security and trustworthiness requirements commensurate with the criticality of the component and in light of all-source threat information. The Science and Technology strategic element is envisioned to identify best-of-breed tools and techniques, and to provide such tools and expert support to acquisition programs. It should also provide incentives to and coordinate with industry to invest in developing standards and tools for vulnerability prevention, detection, and mitigation. Outreach efforts focus on standards activities and seek to build a national market for relevant classes of assured products and the tools that support building, testing and certification.

In the opinion of the Task Force, the systems assurance CONOPS, properly resourced, is an effective way for DoD to organize its risk manage processes to address the system assurance problem efficiently and cost effectively. Details of the Systems Assurance CONOPS are described in the recommendations. Success of the program also hinges on a significantly enhanced IC role only touched upon in the CONOPS. This fuller role for Intelligence and its relationship to the CONOPS will be discussed in the recommendations. Therefore, the Task Force endorses the engineering-in-depth concept of the DoD Tiger Team 2006, which will be explained later in this report.

Other Efforts within DoD and Government Agencies

Aside from the efforts of the DoD Software Assurance Tiger Team to strengthen DoD policies on software assurance, DoDs broader involvement in software assurance remains fragmented. The focus is on defense-in-depth, the Information Assurance Vulnerability Alert (IAVA) software patching process, and the decisions of individual program managers within the services to stipulate software assurance contract requirements--code review, need for security clearances for custom code developers, vetting of COTS vendors, etc. The DoD has recently issued policy on software assurance procedures required when contracting for managed enterprise services, but there is no overarching policy or guidance in place to support program managers in managing these risks.

Most other government agencies simply do not have DoDs focus on software assurance, both because they have less at stake and far fewer resources. In 2006, the Committee on National Security Systems sponsored a Global Information Technology Working Group to propose ways to mitigate risks of supply chain attacks on software and hardware used within the National Security Systems. The report, entitled "Framework for Lifecycle Risk Mitigation for National Security Systems in the Era of Globalization," was released in May of 2006. The Working Group found the risk to national security systems from globalization to be increasing rapidly and judged DoDs defense-in-depth approach as no longer sufficient given the new challenges of globalization. Many of the Group's recommendations for the DoD parallel those of the Tiger Team, whose members participated in the effort. Some of the Working Group's key recommendations were:

- Adopt a "defense in breadth" policy to encourage and even mandate the use of a broad range of security policies and procedures that mitigate the risks from globalization.

- Identify standards and best practices for establishing security requirements and better assured IT products and services by bringing together key elements of the U.S. Government, industry, and academia.

- Create a national clearinghouse for all-source information about suppliers of software and hardware, to improve the government's understanding of any risks associated with the use of specific suppliers in acquisitions.

- Advance U.S. capabilities to test and evaluate IT products to determine proper assurance levels, including the creation of a national focal point to coordinate test and evaluation advancements across government, industry and academia.

The Working Group was co-chaired by DoD and DHS. In addition to DoD, DHS is one of the few Federal agencies actively engaged in seeking broad-ranging mitigations to the software assurance risk posed by globalization. DHS efforts in this area, which are coordinated with DoD/NII, focus on working with the software development industry and academia to raise the bar on acceptable levels of unintentional vulnerabilities routinely written into COTS software. DHS is looking at how to encourage universities to inculcate the importance of developing secure software into the curricula. DHS is also sponsoring the creation of an acquisition management guide for software assurance, aimed at assisting government program managers and contracting officers with incorporating requirements for software assurance within their Request for Proposals (RFP) for custom code development and COTS purchases and integration.

Supplier Trustworthiness Considerations

FINDING: DoD does not consistently or adequately analyze and incorporate into its acquisition decisions the supply chain threat information that is available.

It is not currently DoD policy to require any program, even those deemed critical by dint of a Mission Assurance Category I status, to conduct a counterintelligence review of its major suppliers, unless classified information is involved. Indeed, the current DoD acquisition model for procurement of hardware, software and systems does not require a counterintelligence review of the proposed supply chain for any acquisition. Some programs do conduct such analysis of their own volition, but such analyses do not follow a particular methodology and typically focus on quality and availability of product. Even if such analyses are performed, there is no accepted process for excluding the supplies that fail the evaluation.

A few programs, such as FCS, do require that the prime contractor make an effort to review information on its subcontractors. Even in the unusual circumstances where such reviews are conducted, they do not involve the use of all-source analysis or a methodology to conduct the review. Up until now, most programs conduct no systematic counter-intelligence review of the supply chain, and programs could not request assistance from the ICARC or even learn what information was already available there.

Recently, DoD has established a nascent presence in ICARC. However, programs are generally unaware of that presence, and are not required to utilize it. In any event, the current presence in ICARC of two DoD officers would be insufficient to review even the most critical systems.

Supplier trustworthiness enters into existing DoD acquisition processes primarily for protection of classified information and for research technology protection. The notion of supplier assurance included within the Systems Assurance CONOPS is only incidentally addressed by the policies for protection of classified or research technology protection. Programs protected by these policies may still have flagrant failures of supplier assurance, both because those policies that do exist are often not adhered to, and because the policies are inadequate to the task.

Protection of classified information is addressed through the National Industrial Security Program (NISP) and Operating Manual (NISPOM) and the associated personnel security processes. Properly applied, these policies ensure that classified information is only distributed to suppliers that are not under foreign ownership control or influence, or that any preexisting foreign ownership or influence is adequately mitigated through proxies and special security agreements, and that these suppliers have adequate security practices and vetted personnel to store and properly use the classified information. Research technology protection is implemented through policies on program protection, the use of anti-tamper and the provision of counterintelligence support to prevent technology espionage.

These traditional counterintelligence and security activities are principally designed to protect classified information and technology from espionage as it enters the U.S. industrial base or is used in an operational setting. The supplier assurance element of system assurance is concerned with the protection of DoD systems and networks from corruption through the incorporation of tainted components from the supply chain. These supply chain risks exist regardless of whether classified information is distributed to the supplier, and even if DoD research technology is not involved. From a systems assurance perspective, supplier trustworthiness considers adversarial control and influence of the business or engineering processes of the supplier, as well as the strength of the business and engineering processes from outside penetration.

Other National Security Acquisition Models

The intelligence community has a more rigorous process to conduct all-source analysis of its suppliers, aimed at discovering counterintelligence threats, as well as companies with criminal, terrorist, fraud or financial issues. The process is used in some parts of the intelligence community to examine all primes awarded contracts, whether classified or not, as well as any subs doing classified work. Other parts of the intelligence community use the process more selectively, targeting high risk areas, and particularly software and hardware buys. Some parts of the intelligence community require that all custom code be written by U.S. citizens, and even only by cleared individuals. However, the intelligence community does not routinely require its contractors to develop software and systems through a process such as that proposed by the DoD Tiger Team.

Detecting Vulnerabilities

FINDING: The growing complexity of the microelectronics and software within its critical systems and networks makes DoDs current test and evaluation capabilities unequal to the task of discovering unintentional vulnerabilities, let alone malicious constructs.

Vulnerabilities are weaknesses in software. As with other disciplines of engineering, such weaknesses result from design or implementation of the system. Also, such weaknesses can be unavoidable, accidental, or intentionally introduced, and intentionally introduced vulnerabilities are expected to be well concealed. For large scale software, most vulnerability detection techniques struggle toward high reliability detection of accidental weaknesses, and are far more likely to find accidentally introduced weaknesses, where there is no effort at obfuscation. In this context, finding deeply hidden functionality such as well concealed malicious functionality is a much harder problem.

For most software systems, because the number of combinations of internal states is practically infinite, it is not possible to effectively uncover such hidden functionality through blind testing. Discovering such hidden functionality requires thorough analysis by human experts working with powerful tools in processes not conducive to automation, and this level of reverse engineering and thorough manual analysis is rarely performed except for things like nuclear command and control systems. Moreover, the art of finding hidden functionality is interleaved in a number of ways with the art of hiding functionality, so it is not a skill easily taught broadly without complication, lest too many people are in the art of hiding functionality.

FINDING: The problem of detecting vulnerabilities is deeply complex and there is no silver bullet on the horizon.

Source Code Analysis Tools

One class of source code analysis tools, model checking, has improved dramatically over recent years. Research has produced tremendous progress in software analysis tools and techniques for verifying safety and security of software prior to software deployment. However, it is also important to understand the limits of what such techniques might ever be capable of achieving. Progress over recent years includes lowering false positives from hundreds of false positives per true positive to a level where false positives and true positives are roughly equal, while dramatically increasing the number of bugs found and breadth of types of bugs found. Some of the current tools can find the vast majority of bugs in immature codebases. Perhaps more importantly, for very mature codebases, including large codebases with millions of lines of code, such tools can even quickly and efficiently find prior to deployment roughly a third of the bugs ever publicly found after deployment. Other tools scale to codebases including tens of millions of lines of code. However, it is not clear that any tool will ever be able to find more than half the bugs without improving languages and tools for specifying or expressing desired behavior through means more clear than source code.

Software has continued growing exponentially in size and complexity, further complicating such analysis. Massive growth of the codebases for various operating systems and applications was described earlier. Some operating systems and critical applications such as databases already have codebases including tens of millions of lines of code.

Moreover, the sophistication of the United States potential adversaries has also grown over these years, and most of the newly improved source code analysis tools are publicly available to potential adversaries. In this context, with challenges growing exponentially harder as there is continuous, strong progress in the performance of model checking tools, there is no consensus on whether gains or losses are being made on this problem.

Binary Code Analysis Tools

Because what you see in source code is not always what you execute in binary machine code, analysis of source code is not sufficient. The notion of Binary Analysis is often associated with "Reverse Engineering" which Chikosfy and Cross defined as "the process of analyzing a subject system to create representations of the system at a higher level of abstraction." This type of analysis is utilized whenever software is under scrutiny and source code is not provided, or the source code cannot be trusted. It is often used to gain insight into code of questionable pedigree or quality in a search for flaws or malicious intent or -- in the context of modernization activities -- where the source code has been lost or corrupted.

Binary code analysis tools have been improving. However, binary code analysis does not yet scale to support analysis of most software systems. Given the scalability of model checking and its relative effectiveness in finding accidental vulnerabilities in large scale software systems, automation of reverse engineering of models for model checking from binaries seems to be a promising area for research progress. However, even if that automation of reverse engineering could scale to the scale of current model checking, the effectiveness of such techniques would still be limited to accidental vulnerabilities. If the number of accidental vulnerabilities can be dramatically reduced, the opportunity for adversaries to intentionally introduce vulnerabilities—masquerading as non-attributable accidental vulnerabilities—could be dramatically reduced. Moreover, automation of such reverse engineering may also lend itself to more powerful analysis techniques, such as theorem proving, that are currently less scalable but more effective in detecting hidden functionality in the very small software systems that are currently tractable.

The broad activity of Binary Analysis falls into three basic sub-categories: static analysis, dynamic analysis, and hybrids. Static analysis typically involves taking an executable artifact (e.g., an *.exe file) and disassembling or decompiling the binary to produce native assembly code or a higher level symbolic language for an analyst to review. Dynamic Analysis typically involves executing the binary in a closed or safe environment and observing its behavior for a period of time to establish some level of confidence in its normal behavior and identify any abnormal behavior. Hybrids often involve reverse engineering the binary to produce a higher level symbolic language for mechanical reasoning, and human analyst review to produce the test cases for dynamic analyses most likely to uncover hidden functionality. Moreover, it often makes sense for this to be an iterative process, as those dynamic analyses may create structures meriting more thorough static analysis. This is true for several reasons. Data can generate code, code can be metamorphic and polymorphic, and many software systems tested are built in a manner that residuals of one test vector may interact with the behavior of another test vector. In this context, hybrid iteration between static and dynamic analyses is often required to uncover deeply buried functionality.

Static Binary Analysis

Experienced analysts look for clues in the disassembled code that allow for conclusions about the behavior of the binary. There are a number of mature disassembly tools that are available today (e.g., IDA Pro) that support this type of analysis. Depending on the application, an analyst may be looking for general functionality, potential vulnerabilities, or nefarious behavior where security is a concern. This type of analysis is very laborious and does not currently scale well, as analyzing low level assembler code (the result of disassembly) is a very time consuming, tedious task that requires special expertise that is not widely available.

To address this issue, there is emerging work to address the scalability issue that focuses on a disassembler providing an "intermediate representation" of binary functionality that can be further analyzed by other tools that are specialized in identifying specific nefarious behavior or vulnerabilities. Examples include Grammatech's CodeSurfer for x86, the Software Engineering Institute's Function Extraction Program, or work within KDMs proposed Ecosystem. Specific techniques include Affine Relational Analysis (ARA), Value Set Analysis (VSA), and Aggregate Structure Identification (ASI). However, although some of these techniques offer fidelity conducive to model checking, techniques with such fidelity are not known for scaling beyond executable sizes involving hundreds of kilobytes, let alone today's software systems of tens of millions of megabytes, or tomorrows systems of hundreds of millions of megabytes.

However, the stunning distance that needs to be covered should be noted in context; the investments needed to produce such high fidelity techniques were small, made only within the last few years, and are already beginning to bear substantial fruit. In this context, continued and substantially amplified investments in this area seem not only appropriate, but also potentially critically important given the fidelity they may offer in helping automate the process of revealing hidden functionality.

Dynamic Binary Analysis

Dynamic Analysis involves analysis of the software or system in its run-time environment and is often referred to as "black box testing," as analysis is done from an external perspective. Typically, tools such as "debuggers" and "fuzzers" are used. Debuggers are programs used to debug other programs and provide the ability to trace code execution. Fuzzers are tools that supply permuted data to program inputs to exercise the code. Dynamic Analysis tools typically check for known security vulnerabilities, configuration issues, run-time exceptions, memory leaks/corruptions, or other suspicious Input/Output (I/O) activity. Most of these techniques are targeted at pre-production/integration testing and system-level virus and vulnerability scanning to protect against known software vulnerabilities and malicious code signatures, and to enforce security policies.

Additionally, there are a number of mature tools that provide the ability to instrument the execution environment (e.g., Holedek) and that look for specific vulnerabilities (e.g., International Business Machines (IBM) Purifies for memory leaks). Such tools can provide insight into existing vulnerabilities due to poor implementation but are hit or miss with respect to identifying code of malicious intent, as the triggers for such intent must be fired while the code is in its monitored state.

Given that such tools do not consistently identify malicious code, they are often coupled with static analysis for generation of test cases to drive such binary analysis. In this context, static analysis can identify edge cases, corner cases, and test values relative to constants and decision points in the code, and dynamic analysis can map the state of such behavior in fuller context than other forms of static analysis.

Government Access to Source Code

The tools discussed above are primarily useful during code development. The question naturally arises about their utility for the purposes of acceptance testing or certification of product, which could either be done by an independent laboratory or by the U.S. government itself. It is tempting to consider having the government take the source code of a commercial product and run its own vulnerability assessment tools against it. However, this is problematic – if not impossible - for vendors to support for multiple reasons, including in particular the cost of supporting this testing and the risk to the vendor's intellectual property.

Many vendors' licensing agreements explicitly prohibit decompilation or reverse engineering of products except under limited circumstances. There are tools available that can decompile products for purposes of finding security vulnerabilities. Given that license agreements typically do not allow source access and prohibit most decompilation, the U.S. government would need to clarify the protections it would provide to industry, both in the use of their source code and in the information it derives from that code. This issue is complicated by the lack of governmental incentives to protect intellectual property and the lack of a practical remedy for the vendor if their source code is lost or compromised.

In a global environment some vendors might well decide that, if they release source code to one government, they should also release to others. Indeed, one major vendor has taken this approach.[24] While some governments would report vulnerabilities that were found, others might prefer to use that knowledge for their own purposes.

---

[24] See http://www.microsoft.com/industry/government/programs/gsp.mspx.

There is already a burgeoning trade in third parties selling information on non-public security vulnerabilities by subscription. In some cases, the information may be sold to bad actors, while in other cases, the information is sold to firms with arguably high concerns over their security, such as financial institutions. As DoD moves to apply testing mechanisms to commercial software, it needs to consider what policies will be applied to notification of the vendor, and how the information will be disseminated prior to the issue being fixed, or even after it is fixed.

Aside from the legal and ethical issues associated with government access of source code, there are multiple technical and cost challenges associated with using third-party automated tools against a vendor code base. The high rate of false positives on many vulnerability finding tools would burden the vendor with costly support to a government team that would have little in-depth knowledge of the code being examined. Moreover, in the case of large programs there is the likelihood that the tools would not scale to handle the vendor's code base.

Best Practices for Code Development

While there is no silver bullet that will make software more secure, this section summarizes a set of process improvements that are known to reduce vulnerabilities in software. The recommended process improvements address four phases: design, coding, testing and response. There is no reliance on a spiral, agile or waterfall development model.

This section does not address the need for training of development personnel, but it is assumed that all engineering staff will attend ongoing security training. Few schools teach security from the perspective of reducing vulnerability to attack and increasing assurance, so industry must fill that void.

Design

Design is the critical first stage in building secure software. Unfortunately there are few general models or scalable tools that help deliver secure designs for COTS software. However, the process of hazard analysis, sometimes referred to as threat modeling, is useful for understanding the threats posed to software and the errors that can lead to successful attacks. The threat modeling process, defined in[25] calls for structured decomposition of the application into the following elements: processes, data flows, data stores, external entities and trust boundaries. Element types can be attacked in specific ways – for example a data store (such as a file or a database) can be tampered with, denied service, or viewed. The aggregate of all elements within the application and the different types of attacks that can be launched against each element comprises a list of potential threats (or hazards) to the system. Each threat type can be mitigated in well-understood ways with security features. For example tampering can be mitigated with integrity technologies such as permissions and message authentication codes or with attention to assurance techniques.

---

[25] Howard and Lipner 2006

Another important design concept is the understanding of how exposed the application is to attack. The most extreme case is unauthenticated and remotely accessible network end-points opened by a privileged process that runs by default. Information about exposure of an application can be derived from the threat model. The goal of any product development team over time should be to reduce the product's exposure.

Coding

A review of the history of reported software vulnerabilities over the last five years indicates that most security bugs that pose material risk to software users are implementation flaws. In general, these flaws could have been prevented during the development phase. The following paragraphs summarize some key techniques that have proven effective for improving software security and removing vulnerabilities during implementation.

The compilers that support the development of software in higher-level languages, such as C and C++, must add defenses to the generated code. At a minimum, these defenses should include detection of stack-based buffer overruns. While such defenses do not solve the buffer overrun problem, they do add a hurdle for attackers to jump over.

Buffer manipulation functions in the C and C++ languages (for example: strcpy, strncpy, strcat, strncat, and sprint) have proven to be a continuing source of software vulnerabilities for all vendors. These dangerous functions should be removed over time and new code should not use these dangerous functions; more robust and predictable functions should be used instead, such as StrSafe or Safe CRT.[26]

Regardless of the programming language used, static analysis tools should be used to detect some classes of coding errors. These tools should be run regularly and bugs triaged and fixed. It is also beneficial to use multiple tools – different analysis techniques detect different classes of bugs. While some static analysis tools have evidenced high "false alarm rates" – requiring an unacceptable level of human review of erroneous reports – the market is producing additional static analysis tools every year and available tools have proven effective at analyzing even large and complex commercial software.

There are two major classes of analysis tools – those that inspect source code, and those that inspect the compiled executable code. Both have value in that they find different classes of bugs. Software developers need to move towards using both types of tools. However, developers should exercise caution in relying heavily on static analysis tools. Tools can help scale the security analysis process in that they are rapid, but they are no replacement for skilled human inspection. In addition, some tools that simply scan for certain patterns in code are a waste of time, because

---

[26] Howard 2002, Lovell 2005

their false alarm rates are very high. Most of these tools simply look for bad coding constructs without regard for the nature of the data or any other contextual consideration.

Software engineering organizations should consider external design and code review by code-security analysts. Review by another set of knowledgeable engineers – either inside or outside of the organization developing the software – can be effective. However, the vast complexity of much commercial software is such that it could take months or even years to understand.

Testing

"Fuzz testing" can be very effective at finding security bugs. If an application processes complex data structures or network traffic, then it should be "fuzzed." Fuzzing means building deliberately malformed data and having the application consume the data. If the application fails, the developer should review the failure to identify the error and fix it if appropriate. There are two major ways to build malformed data: the first is to randomly corrupt well-formed data and the second is to intelligently corrupt well-formed data. The latter approach means that the tool performing the fuzz test must understand the protocol being fuzzed, which makes the tool more complex and also results in more effective testing. Tools are critical for fuzz testing to be successful because massive quantities of malformed data must be created. A typical fuzz testing process requires that a minimum of 100,000 malformed files be generated for each file format consumed by the application under test.

A benefit of fuzz testing is that once the test harness is built, there is no need for human intervention other than to review software failures. Effective fuzz testing requires an understanding of code coverage. If a fuzz testing process only exercises 25 percent of the data-parsing code, then the fuzz testing process is ineffective. It is imperative that the tests touch almost all, if not 100 percent, of the code.

Another form of security testing is penetration testing in which skilled attackers conduct a directed search for vulnerabilities. Such testing is useful to a point, but is expensive, yields relatively few bugs, and requires a great deal of skill and experience to perform. With that said, the bug quality from penetration testing can be very high if the team performing the review is highly skilled. If the product under review is to be used in sensitive environments or used by large number of customers, then a penetration test might be beneficial.

Security Response Processes

As the preceding paragraphs have repeatedly made clear, there is no silver bullet for achieving secure COTS software. As a result, vulnerabilities will continue to be discovered in fielded software for the indefinite future. Many COTS vendors have recognized this fact and integrated a security response phase into their software development process. Part of the role of a security response process is to accept reports of newly discovered vulnerabilities, investigate them, and act to protect users of the vulnerable software. This action may include maintaining communication with the finder of a product vulnerability, releasing a security patch or update, releasing an advisory with information for users, and providing information to suppliers of anti-malware or intrusion prevention/detection products that customers may use to block exploitation of a vulnerability. By maintaining 24x7 response capabilities with the maturity to manage complex relationships with security researchers, vendors can promote responsible vulnerability disclosure. This enables vendors to release patches before exploits can be launched and reduces the impact of "zero-day" attacks.

The impact of a security response process extends beyond its immediate role of providing users with fixes and mitigations for newly discovered vulnerabilities. A well-organized security response process will identify new classes of security vulnerabilities and ensure that design guidance, development tools, and testing practices are updated to prevent the new class of vulnerability from being introduced into future software releases. Such a process will also incorporate a search for related vulnerabilities in fielded software so that users of the software can eliminate all similar problems by applying a single security update.

Software Assurance in Computer Science Education

One of the root causes of poor software assurance is the current culture of software development. Developers are, in general, not indoctrinated in either the techniques or the values of secure coding practice in their degree programs. As a result, many commercial software vendors have to spend extraordinary time and money educating otherwise bright developers on the principles and practice of secure coding basics as well as why it matters. It may be relatively easy to teach developers how to validate input, but it is hard to convince them that they need to do this properly every time. Vendors will always need and want to indoctrinate their employees in vendor-specific coding practices, but they should not need to be teaching the basics of good, defensible programming.

FINDING: The academic curriculum and courseware in computer science does not adequately teach software developers to design, develop and test with a defensive mindset, nor is there adequate training in assurance and security techniques.

University programs must change so that businesses are not teaching remedial secure coding practice. While some computer science curricula include a security module, it is typically focused on the basics of security features and functions (e.g., authentication, access control, and audit), rather than on the overall practice of secure development.

Most developers do not have a defensive mind set (e.g., "first, assume an attacker…"), nor do they follow principles of building functionality to do what is intended and explicitly prevent or prohibit what is not intended. In other words, developers often explicitly want to allow functionality to be extensible – at some point in the future - without considering the security implications.

Ideally, secure development practice would be embedded within every university class, so that the principles of secure coding and security-oriented development:

- Are seen as part and parcel of any development effort,
- Are emphasized repeatedly, and
- And are more likely to stick as both a cultural value and as practical knowledge.

To incorporate the issue of malicious coding or "supply chain corruption" into this, educational improvements should include both secure coding and, specifically, defensive coding.

Defensive coding should mean that students' code is actively attacked as part of classroom exercises. For example, a computer science professor at one university has red teams and blue teams attack one another's code as part of each homework assignment, and the defensibility of code is part of a student's grade. A subtle but important aspect of this could be assignments to subtly corrupt software so that it does something other than what it is supposed to do without that code being easily detected. Teaching students that this can happen and how to look for it is another way to change the mindset of people who write code.

Universities (with some notable exceptions) have been resistant to changing their curricula to emphasize secure coding practice. Several large vendors have attempted to influence various accreditation bodies (e.g., the Association of Computing Machinery (ACM), and Institute of Electronics and Electrical Engineering (IEEE)) to address this issue. Most universities have, to date, not changed their curricula.

Conclusion

All of the considerations just listed seem to point to an intractable problem. The Nation's defense is dependent upon software that is growing exponentially in size and complexity, and an increasing percentage of this software is being written off-shore in easy reach of potential adversaries. That software presents a tempting target for a nation-state adversary. Malicious code could be introduced inexpensively, would be almost impossible to detect, and could be used later to get access to defense systems in order to deny service, steal information, or to modify critical data. Even if the malware were to be discovered, attribution and intent would be difficult to prove, so the risk of detection for the attacker would be small.

Against this backdrop of potential disaster, practical experience and belief paint a picture of aggravating and continuous software problems, but not ones that are lethal. However, there are some systems on which, to varying degrees, life depends (e.g., power, health). In this sense, DoD systems are among the most critical because their national security mission is often measured in fatalities, and failures that would be innocuous in another context can be lethal and lead to mission failure.

If the attacker cannot be deterred and its malware cannot be found, what is to be done to provide assurance that DoD software will perform in mission-critical situations? First, there never will be an absolute guarantee. But as in many issues in defense planning, software assurance comes down to intelligent risk management. The risk of vulnerable software can be managed through a suite of processes and mitigation strategies detailed in the Task Force recommendations, and this risk can be weighed against the attractive economics and enhanced capabilities of mass-produced, international software.

RECOMMENDATIONS

DoD Procurement of COTS and Off-Shore Software

RECOMMENDATION: DoD should continue to procure from, encourage and leverage the largest possible global competitive market place consistent with national security.

The reasons underlying DoD use of commercial software described in the 1999 DSB report "Globalization and Security" remain as true today as when that report was written. Much functionality required by DoD for its warfighting mission can be more effectively obtained from commercial industry from the perspective of cost, schedule and performance. In fact, failure to use commercial solutions for much functionality would be cost prohibitive, even assuming that functionality could be produced by the traditional DoD industrial base.

It is a reality of the market place that the supply for IT will globalize. Competitive forces drive companies to seek technical talent, cost savings, new product markets, and 24-hour production cycles. Such globalization appears to lead to higher degrees of competitiveness, leading to better products and services at lower prices. Thus the DoD must necessarily continue to purchase software written offshore.

DoD has greatly benefited from these circumstances. In addition to cost savings – DoD shares in costs spread over the entire market of consumers – product maintenance, technology refresh, and market innovation all benefit DoD. So long as DoD is not put in a position of being denied the best technology (reverse-International Traffic in Arms Regulations (ITAR)) through the loss of U.S. industry competitiveness, and so long as commercial products destined for DoD systems are not used inconsistently with the levels of assurance that can reasonably be expected from those products, DoDs agile use of commercial technology in support of the warfighter is another avenue of competitive advantage.

Increase U.S. Insight into Capabilities and Intentions of Adversaries

RECOMMENDATION: The IC should be tasked to collect and disseminate intelligence regarding the intents and capabilities of adversaries, particularly nation-state adversaries, to attack and subvert DoD systems and networks through supply chain exploitations, or through other sophisticated techniques.

Based upon the staggering consequences of defensive failure, a path forward toward an actionable understanding of nation-state adversarial intents and capabilities must be identified, resourced and executed. Components of this path should include:

Recommendations

- The IC must be tasked and resourced to aggressively collect, analyze and report on the capabilities, methods, intentions, and targets of foreign offensive services targeting the U.S. This insight must be in sufficient detail to be actionable by U.S. defensive players.

- Change the culture, through education, to take into account the real threat coming from the capabilities of nation-state adversaries. DoD can no longer plan and operate as if its mission-critical systems were functioning in a cyber environment free of such adversaries.

- The national security community needs to foster improved analytical capabilities that can discern the intent of broad patterns of intrusions and organize countermeasures that blunt the overall efforts in addition to incident by incident response efforts. Of course, sophisticated adversaries can attempt to compromise people with access to IT systems, a threat that requires its own rigorous protections.

All indications are that targeted attacks against defense systems will increase in sophistication and stealth. Federal enterprises in general need to develop better detection and analytical capabilities to more rapidly identify, assess and respond to stealthy attacks. DoD needs to understand both how attacks are occurring and the types of data sets being targeted.

RECOMMENDATION: DoD should increase relevant knowledge and awareness among its cyber-defense and acquisition communities of the capabilities and intents of nation-state adversaries.

It is vital that senior leaders, acquisition officials, program management staff and trusted industry partners understand the reality of the threat, vulnerability and consequence of the nation-state adversary targeting DoD software systems. In order to ensure their understanding, barriers to sharing threat information from the offensive operations community must be identified and changed. The vehicles for communicating threat to the acquisition community (e.g., System Threat Assessment Report (STAR)) have to be evaluated for adequacy. In addition, the changes described here will require a cultural change that must leverage education and training of senior executives and the acquisition workforce.

The Task Force earlier described the natural hesitance to share sensitive operational information. However, it is necessary that the IC develop policies to mandate the timely cleansing and communication of threat information about the capabilities and intents of nation-state adversaries to senior leadership within the acquisition and defensive operations communities. In addition to leadership, mechanisms will have to be developed to enable technical experts within the defensive operations and acquisition communities to understand what the technical threat is.

The STAR is generally tailored to describe the threat to the intended system in the operational setting, rather than from a full lifecycle perspective, and may not fully consider the sophisticated capabilities of the nation-state adversary. Moreover, the STAR is generally not required before Milestone B of the DoD program structure. The ability of the nation-state adversary to attack technology through lifecycle exploit should be described to the Project Management Office earlier within the lifecycle.

System Engineering and Architecture for Assurance

RECOMMENDATION: DoD should allocate assurance resources among acquisition programs at the architecture level based upon mission impact of system failure.

Classic risk management principles require starting by identifying assets to be protected, determining the importance of those assets, determining threats to those assets, and then mitigating those threats. In considering the importance of an asset, it is important to consider not just the asset alone, but also dependencies (e.g., which other assets will fail if this one fails).

DoD cannot cost effectively achieve a uniformly high degree of assurance for all the functionality it uses across its many and varied mission activities. Allocating criticality of function levies a requirement for assurance of that function, and also of those functions that defend it. Systems identified as critical must then allocate criticality at the sub-system and assembly level.

To properly allocate scarce assurance resources, DoD must allocate criticality at the system-of-systems and enterprise architecture level. This analysis should occur early within the system life-cycle, and should render a prioritization decision no later than Acquisition Milestone A, to allow programs of record to respond appropriately to their criticality.

DoD categorizes systems by mission assurance category and confidentiality level for information assurance purposes. If these categorizations are made to link the system's criticality to its role within the DoD warfighting mission from a system-of-systems or architecture perspective, it would constitute an architecture-level allocation of criticality. Currently, allocation of mission assurance category and confidentiality level does not appear to systematically drive an allocation of the requirement for assurance at the subsystem, assembly, and component level.

While DoD determines how to identify its system criticality, the Task Force recommends that the USD(AT&L) and the ASD(NII)/CIO empanel a stakeholder group composed of service representatives, the Joint Staff and OSD to identify a short list of critical systems to which the CONOPS would be sensibly applied in light of current acquisition lifecycle exigencies.

Recommendations

The CONOPS is not intended to be applied to legacy system functionality, but will apply to new systems and modifications to critical legacy systems. If a software threat is discovered in an acquisition program, for example because of a previously undetected supply chain threat, then vulnerability management systems should be applied to determine the vulnerability in legacy systems.

Currently, systems engineering processes do not adequately identify the requirement for assurance, do not adequately engineer for the system risks resulting from the composition of individual components with software and hardware assurance risks, and do not adequately manage system assurance risks. Even within a DoD Critical System, not all functions will have the same criticality to the mission-worthiness of the system.

For critical systems, DoD should implement enhancements to systems engineering processes to meet systems assurance requirements. These assurance requirements will be established for critical-system acquisition programs through a prioritization process involving the following factors:

- Assessing the vulnerabilities of alternative system designs from a failure modes perspective.
- Minimizing the number and criticality of components requiring greater assurance.
- Acquiring these components from assured suppliers and managing the risks inherent in the use of less assured products.
- Defending intrinsically critical functionality with mechanisms that are considered to have the same criticality as the components they are defending.

The DoD Tiger Team has termed this system engineering process "Engineering in Depth."

For each critical system or component systems designers should ask: "What is the worst thing that a malevolent insider could do"; and then ask, "How could such an action be mitigated"? Red Teams of specialized personnel to address system, and system-of-system, attack scenarios should be employed very early in the architecture processes and continue throughout system development. Red teams need not be large, but do require a few key people with substantial knowledge of computer science and development, system vulnerabilities, and potential attack tools and methods.

Mission and information assurance cannot be added as an appliqué late in the process, or after system development. It must be an integral part of development to provide graceful degradation, isolation, multi-patching, replaceable modules, and other techniques to manage risk. Critical systems must be able to recognize an attack or undesired state change, resist by preventing further damage, and recover such that critical functions can be performed at critical times. Creating these

Eliminate Excess Functionality in Mission-Critical Components

Each new capability introduces opportunity for new vulnerabilities and excess capability means unnecessary vulnerability. This means that there is a potential downside to every new capability incorporated. The addition of a new side door to a relatively secure building means added convenience, but introduces the requirement for additional physical security -- added functionality, added vulnerability. It is necessary to question closely the need for each additional capability. In this context, accepting features of no use to the DoD with the risk of adding vulnerability seems an inappropriate course.

Frequently, however, the acquisition system is biased in favor of excess functionality, without factoring in the risk that such unneeded functionality brings. In a competitive evaluation, when all systems meet the parameters for needed functionality and cost, those extras can be the tie breaker ...in the wrong direction. Even the requirements system in advance of actual acquisition happily compiles a long list of "nice-to-haves" —really a list of dangerous ingredients. The remedies to this are good system design and careful evaluation of the cost of adequately defending mission-critical functionality from "nice-to-have" functionality.

Improve the Quality of DoD Software

The DoD can effectively raise the "signal-to-noise ratio" against software attacks by raising the overall quality of the software it acquires. If there were fewer unintentional bugs in software, the visibility of poorly written malware might be increased. While general improvements in information assurance will not, per se, prevent a determined attacker from corrupting the software supply chain, there are several compelling benefits in improving the overall assurance/security-worthiness of COTS:

- A sophisticated adversary would have to work harder to introduce an exploitable vulnerability instead, as is currently the case, of relying upon the plausible deniability of a common programming error to avoid attribution of malicious intent. Furthermore, a sophisticated adversary would have less confidence that its malware would remain undetected, invisible in a world containing far fewer distracting vulnerabilities. That uncertainty could be a deterrent in itself. Any attacker now has a plethora of exploitable defects to choose from.

- Eliminating common, easily-exploitable defects in software will, as one software industry executive has put it, "help get 16-year-olds out of DoD networks." The "signal to noise" ratio for defenders improves as easy attacks are rendered more difficult and thus, more scarce. By reducing the resources that must be expended to combat opportunistic attacks, redirecting saved resources can be used to prevent, detect, and respond to more sophisticated adversaries.

Improve Tools and Technologies for Assurance

Improve Trusted Computing Group Technologies

The Trusted Computing Group is a not-for-profit organization formed to develop, define, and promote open standards for hardware-enabled trusted computing and security technologies, including hardware building blocks and software interfaces, across multiple platforms, peripherals, and devices. TCG specifications claim to enable more secure computing environments without compromising functional integrity, privacy, or individual rights. The primary goal is to help users protect their information assets (data, passwords, keys, etc.) from compromise due to external software attack and physical theft. The Trusted Computing Group was initially formed by personal computer manufacturers, suppliers and software companies AMD, Hewlett-Packard, IBM, Infineon, Intel Corporation, Microsoft, and Sun Microsystems, Inc. Today it has over 120 members from across computing, including component vendors, software developers, systems vendors and network and infrastructure companies.

The centerpiece of the Trusted Computing Group initiatives is the Trusted Platform Module (TPM): a semiconductor chip or chipset that is embedded on the motherboard of personal computers and other computing platforms and components. Trusted Computing Group functionality has also been integrated directly into certain mass-market chipsets. Since each TPM chip or chipset is unique to a particular device, it establishes machine identity that cannot be forged and binds keys to a specific machine state. Therefore, it can be used to authenticate a hardware device, and can be used to verify that the system seeking access is the expected system.

TCG-compliant modules and components are designed to prevent software attack and they are capable of preserving software assurance once an assurance level has been established. However, the TCG specifications are silent on determination of initial assurance levels. NSA and others have identified a number of improvements and complementary practices that would strengthen TCG-compliant systems, including privacy-preserving attestation, virtualization, and architectures that provide richer software assurance measurement and monitoring capabilities.

Given that vulnerability detection can never be perfectly reliable, it seems necessary to bound the harm that can be done with hidden functionality escaping detection. A variety of TCG technologies lend themselves to helping enforce separation of information so that processes from one domain will not harm processes or information in another. However, at present, these protections are not provably strong. For these reasons, the Task Force proposes that the U.S. Government fund unclassified research to analyze and improve the strength of TCG technologies for providing such separation properties. The goal of such a program should be security and virtualization software and chipset design that are formally proven to provide the desired separation properties under all circumstances.

Improve Effectiveness of Common Criteria

Currently, the official DoD-wide evaluation/validation scheme is the National Information Assurance Partnership based upon the Common Criteria. The reality today is that it would be far easier and more effective to improve Common Criteria than to invent a new scheme specific to the DoD or to DHS.

Specifically use of the Common Criteria and NIAP practices should include:

- Assessing vendor software developmental processes and vulnerability mitigation processes to give confidence in their change-management procedures throughout the software products lifecycle.
- Crediting vendors for the effective use of automated vulnerability reduction tools during all areas of the software development lifecycle.
- Rating the effectiveness of the automated vulnerability reduction tools.
- Use of automated tools for vulnerability analysis during EAL4 and below evaluations.
- Timely and cost efficient evaluation/validation schemes reducing artificial artifact creation.
- Policies and processes for sharing of vulnerabilities within the acquisition processes.

The National Information Assurance Partnership (NIAP) Common Criteria Evaluation and Validation Scheme (CCEVS) has begun to develop an evaluation process that appears to meet many of the objective outlined above. The Common Assurance Assessment (CAA) has been developed with the cooperation of the United Kingdom's (U.K.) Common Criteria scheme, and has been used for an initial "alpha test" trial evaluation in the United States. A second trial evaluation, to be conducted by the U.K. scheme, is planned for 2007. CAA is still under development and there are many issues yet to be resolved. However, it holds promise because it focuses on a vendor's development process as actually conducted, and because it examines the actual process and artifacts associated with development rather than artificial documentation written solely for the evaluation process. DoD should continue to develop and evaluate CAA with the aim of using it as a foundation for a new way of gaining insight into the assurance of COTS software.

Commercial development processes are changing. As part of that change, new artifacts are being created (e.g., threat models). To the extent these artifacts are reflective of real software security and are actually produced during development, they can be relied upon in a certification process. However, sometimes the artifacts created may not actually reflect accurately the product. This occurs, for example, when that artifact (e.g., a design) is not followed precisely in the implementation. To the extent the vendor's artifacts are not sufficient -- either because the government wants something the vendor did not create or because the artifact provided is not good evidence -- it is reasonable to assume the government will want some other artifact created. The problem is that any artifact created after the fact did not influence the product and thus meets no business need aside from certification.

It would be beneficial for government and industry to work more closely together to align artifacts created by reasonable development practices and what the government relies upon. Where there is no synergy, there are only two possibilities. First, the industry has created something for its own purposes that the government does not want to use, or second, the government shows a reasonable need for the item and industry creates it solely for the purpose of certification. This becomes a cost of doing business.

Improve Usefulness of Assurance Metrics

There is a natural tension between the U.S. Government's need to know the security-worthiness of what they procure and vendor's need to avoid disclosing particular vulnerabilities. One way to satisfy both needs is to develop a weighted index of the security-worthiness of software. A weighted score could be generated via testing based on some combination of:

- The utility of the tool itself: what vulnerabilities it finds, how well it finds them and how accurately (e.g., some sense of the false positives and false negatives generated by the tool).
- The amount of code coverage of the tool -- how broadly the tool is run against the software, and how broad is the range of vulnerabilities checked.
- The test results against a particular product, weighted for the severity of outstanding (unfixed and/or non-mitigated) issues.

The issue is DoDs desire to know what the vendor knows. Put another way, what DoD should want is the vendor to say: "These are the tools we used; we covered "n" percent of the code base, and here is what we found."

The vendor could use an automated tool of choice (presumably, what is used during development), a laboratory could revalidate the results, and there could be some "rating" of the tool (by NIST) for what it finds and does not find. The goal would be not to disclose the details of specific vulnerabilities found, except to the vendor. The U.S. Government would thereby get an idea of the overall security-quality of software weighted for a number of critical factors: "a goodness meter."

To make this feasible NIAP should develop a ratings methodology for development processes, as well as automated vulnerability-detection tools and a weighted metric (or collection of metrics) to use against commercial software.

Promote the Use of Automated Tools in Product Development

Tools for the detection of vulnerabilities continue to improve and proliferate. Different vendors will choose to use different tools; no one tool is right for every company. The DoD may well have an interest in determining how good the tools are that vendors use, but DoD should not be in the position of dictating particular tool use by vendors. Requiring vendors to validate their products using product X (even if it is done by a lab) effectively means all vendors selling to the government must purchase a product X license or service, because nobody wants to be surprised by the test results from a lab. Creating de facto standards is inappropriate and DoD cannot reasonably pick a "code scan" winner that will be equally deployable, cost effective, useful, scalable, etc. in all environments. Furthermore, different types of tools have different utility to different vendors. A vendor of a protocol-heavy product may derive greater benefit from using a protocol fuzzer than a static analysis tool, for example.

Markets should pick winners; that said, DoD could, in theory, rate vulnerability analysis tools for what they find and how well they find it by using a neutral third party like the National Institute of Standards and Technology (NIST) or a commercial evaluation. Such ratings should clearly reflect the range of size and complexity of the code base where the analysis tool is found effective. This would be a benefit to vendors who want to use such tools and have no idea how good a tool is, or who have limited ability to vet them. However, there is an advantage in having multiple tools on the market. Small, innovative startups need to be able to use these tools, not merely large, established vendors. Having a number of tools on the market at different price points will help.

More Knowledgeable Acquisition of DoD Software

Increase Transparency and Knowledge of Suppliers' Processes

RECOMMENDATION: DoD should implement a scalable supplier assurance process to assure that critical suppliers are trustworthy. DoD should coordinate with the Director of National Intelligence (DNI) to leverage the all-source threat assessment capability resident in the DNIs Intelligence Community Acquisition Risk Center (CARC), possibly shaping this into the foundation of a national supply chain threat assessment capability.

When government agencies contract for software development in response to a specific requirement or statement of work, they should specify the process that the contractor will be required to use, the tests that the contractor will conduct, and documentation and deliverables that the contractor must produce. In contrast, vendors of COTS software apply processes that are dictated by their own perceptions of customer demand and the best ways to meet it. Each software vendor devises its own process in response to its business needs, and there is no requirement for a vendor to make either its process or the artifacts the process produces available to customers or third-party reviewers.

In order to gain meaningful assurance of COTS software, DoD must understand the actual processes used to develop that software. Gaining this insight into vendors' development processes is not the same as mandating a process and verifying that it has been followed – software development processes are core to vendors' success in the worldwide government and commercial market, and any successful vendor would be hamstrung by DoD processes that conflicted with that vendor's ability to be successful in the broader marketplace. No product evaluation regime in effect today provides insight into a vendor's real development processes and their effectiveness at producing secure and trustworthy software – so the software assurance challenge for DoD is to define an evaluation regime that is capable of reviewing vendors' actual development processes and rendering a judgment about their ability to produce assured software.

An evaluation regime that meets DoDs requirement for assured software must have the following properties:

- It must examine code and the processes used to design, implement and test that code, because software vulnerabilities are most often created at the implementation (code) level.

- It must assess the tools vendors use, their effectiveness at finding vulnerabilities, and their integration into vendor development processes, because effective software development processes depend heavily on the use of automated tools to detect and remove vulnerabilities.

- It must include some opportunities to interview development staffs, review their training, and assess their ability to produce assured software, because commercial vendors rely on the competence and commitment of their employees as well as (or, in some cases, instead of) formal processes.

- It must provide an extraordinarily high standard of protection for vendors' intellectual property, because source code and development processes are the crown jewels – along with employees, literally the only assets – of software vendors. Under Common Criteria, which arguably exposes less information to evaluators than the process being envisioned, the contracts that vendors sign with the commercial laboratories of their choice can provide very strong commercial remedies for any mishandling of vendor proprietary information. Any new evaluation scheme should provide more relevant information regarding assurance than the Common Criteria.

- It must include an assessment or scoring mechanism that allows DoD decision-makers to compare very different kinds of products and processes, because different software products must meet different requirements and face different threats and because vendors apply different processes. The only feasible way to provide such a scoring mechanism is to examine a software development process at a fundamental level and evaluate the elements of the process as a whole made up of individual parts.

DoD procurement and security professionals need to better understand and leverage key business drivers to gain the assurance that vendors are applying appropriate security controls over their products. For example, COTS vendors execute control over their source code to ensure that there is a clear chain of custody to protect intellectual property, to ensure that coders do not inadvertently introduce unwanted code elements such as those that may be covered by alternative licensing schemes such as GPL, and to ensure quality control and adherence to their respective development guidelines. Weakness in any one of these three areas could be costly to a vendor, damaging its brand reputation and costing it market share. Ultimately, understanding the processes and controls used by vendors is a sophisticated approach to managing COTS acquisition risks, and can be used in conjunction with such approaches as analyzing the historical security relationships of countries where products are developed or the particular citizenship of those in the software supply chain.

DoD needs to develop a policy requiring a thorough counterintelligence scrub of its key suppliers for critical systems. DoD should subject to an all-source assessment any company conducting classified work for such systems, and any company writing software or selling a COTS product that will be utilized as a critical component in a critical system. At present such analysis is conducted for the IC by the DNIs Community Acquisition Risk Center (CARC), which has some eight years of experience and a demonstrated threat analysis methodology. It is recommended that DoD identify a point of contact within its security/counterintelligence entities to conduct required threat assessments for programs. OSDs Counterintelligence Field Activity already has a nascent presence in the ICARC, as does one or more of the service counterintelligence shops.

Suppliers should be characterized in terms of the direct threat that they present to DoD acquisition programs, which is a requirement consistent with ICARCs current charter. However, suppliers should also be reviewed for any indirect threat they pose through weaknesses in security-related engineering and business practices and procedures. CARC methodologies do not currently encompass all of these issues. All these considerations should enter acquisition through source selection decisions.

Components Should be Supplied by Suppliers of Commensurate Trustworthiness

RECOMMENDATION: DoD acquisition processes should require that products possess assurance matching the criticality of the function delivered.

The critical components of critical systems demand higher levels of assurance. In this context, critical components are software custom code and COTS that an engineering analysis of the system designates as necessary to ensure at least a defined level of minimal performance.

RECOMMENDATION: DoD should require that all custom code written for systems deemed critical be developed by cleared U.S. citizens.

Cleared personnel should also review legacy custom code being incorporated into new critical systems for malicious code or vulnerabilities.

While it is impractical to conduct any significant review of COTS software packages used even in critical systems, it should be a requirement of new systems that such COTS be verified to the extent practicable if it is designated a critical component. The level and extent of verification should be determined by the Program Office as a part of its overall risk mitigation strategy to ensure operational capability of the system.

Acquisition Incentives for Higher Quality Code

The U.S. Government collectively provides a very large market for software and services. The collective buying power of the U.S. Government is such that it can force change on its suppliers to a degree no other market sector can reasonably do, with or without the formal regulation that serves as a baseline for suppliers. The U.S. Government taken as a whole seems unaware or unwilling to use its clout in the marketplace. There are some strong pockets attempting to change this that have met with mixed success. In particular, some efforts have been hampered by a lack of dialogue with industry about what is feasible in terms of product change timeframes. Procurement should be used to push for constructive security change, but it must include sufficient vendor notice and feedback mechanisms, some grandfathering, and should acknowledge product lifecycles (in which major product changes take years, not months to implement).

DoD should alter the purchasing dynamic with the vendor community by insisting on more security-worthy software through their procurement cycle. This should be done by providing a generally uplifting effect on software instead of merely imposing more cost without commensurate benefit or by excluding classes of vendors. Companies will not build a robust, secure version of a product for DoD and a buggy, insecure version of the product for their commercial customers.

RECOMMENDATION: DoD should provide incentives to industry to produce higher quality code.

There will never be a single set of best practices, testing tools, methodologies, or development process that works for all vendors -- nor is that desirable, since innovation requires change. That said, it is reasonable to expect some cohesiveness to formulate around good development practice and to expect industry to do more over time to implement strong development processes that improve security-worthiness and quality.

In the past, market dynamics have not penalized poor quality software, and in some cases, and has even created concerns that such poor software generates a market for maintenance services and security products. It is, of course, not that simple; most vendors provide security patches for free and many security products protect against more than vulnerabilities (e.g., unsafe user behavior such as clicking on attachments from unknown sources). While the popular sentiment is "caveat emptor" (let the buyer beware), the fact is that customers generally have insufficient information about the quality and security-worthiness of commercial software to make fully informed purchasing decisions, or to know what the lifecycle costs of securing the software will be. Several factors work against quality being a measurable factor prior to purchase:

- Many vendors charge for maintenance, which includes both patches and upgrades. Producing patches thus becomes a revenue source for vendors and a reduced (and perverse) incentive to provide stable software. This bundling provides an advantage to customers in that both enhanced functionality and fixes to broken functionality are typically covered by maintenance charges, but it also penalizes customers who want stable software and do not need or want new features or upgrades.

- Many patches, particularly in security, are issued on unpredictable schedules. Customers do not have any way of knowing what their actual patching costs will be. This would include patch analysis, testing, downtime, actual patch application, and recertification of systems. The license cost (including maintenance) is typically a fraction of the overall lifecycle costs from the customer perspective, and the larger portion cannot adequately be factored into purchasing decisions (unless, of course, one opts for software as a service, in which case the vendor assumes more of the risk of unstable software, as the vendor is typically charging the customer a fixed rate to both supply and support the software).

- Many vendors do not have good insight into what producing patches actually costs them in terms of lost developer productivity, opportunity cost (e.g., other features they could have built with developer resource focused on patches), and so on. Also, as noted above, maintenance may be a revenue center for some vendors.

- Industry analyst firms generally do feature-to-feature comparisons of software products but in general do not do cost of ownership comparisons. Doubtless the great difficulty of comparing lifecycle costs is part of this, but the fact that so many analysts are also paid by vendors for consulting probably contributes to their reluctance to do head-to-head cost comparison for fear of alienating their paying (vendor) customers.

- The IT market tends to reward time-to-market for cutting edge technologies, which means the incentive is to skimp on security, testing and other

Recommendations

"quality" portions of development processes in favor of getting product out the door quickly and grabbing market share. Customers who adopt the newest technologies may often suffer through the poor quality phase of the product lifecycle in part because they have no alternatives, and because the cost to switch is too high.

To remedy this, DoD needs to help change the market dynamic through both positive and negative incentives so that they get better quality software, and to make better risk-based and "total cost"-based acquisitions. Note that vendors who build better quality COTS software (for DoD) are unlikely to sell lower quality software to other customer sectors; in fact, it is likely that any positive impact from DoD helping to raise the bar will accrue to other sectors of critical infrastructure.

The pace of introduction of new software applications and new functionality far outpaces traditional product evaluation schemes and system certification and accreditation processes. To ensure that the DoD has access to and understanding of novel, state-of-the-art software it must continue to develop and exploit rapid evaluation processes and techniques. This should not be restricted to applications that might only be used in classified environments but should also be used to evaluate software that might enhance or influence unclassified networks and systems. This rapid evaluation capability, while segmented and separated by various levels of classification, should include where appropriate participation by universities and institutions involved in software development assessment and certification.

Technically, shipping products in a default secure configuration (and making it easy to maintain a reasonably hardened security configuration) is not a quality measure in the sense of defects per thousand lines of code is a quality measure. It does, however, go to cost of security ownership, lifecycle costs and the overall security posture of a product. If a vendor can ship a product in a pre-hardened condition, that saves customers the cost of reading documentation, tweaking multiple controls to lock the product down, and testing that the hardened configuration works in their environment. The cost avoidance to customers by the vendor doing this upfront is tremendous. DoD has already used this successfully in contracting vehicles. The USAF recently required a hardened configuration as part of an enterprise license agreement; they claim this will save the USAF millions over the lifecycle of the contract.

Risk-Based Acquisition

Requiring or dictating a standard development process is not the path of greatest benefit to DoD. Having said that, knowing what vendors do as part of their development process helps DoD evaluate risk. Also, the mere fact of asking what vendors do to engineer security and quality into their lifecycle puts the vendor community on notice that it is important to DoD.

The DoD/DHS software assurance forum has been working on a procurement guide focused on software assurance, which helps procurement officers glean (through a series of questions) what vendors have done (and not done) as part of their secure development process, how they handle vulnerabilities, and so on. Such a document, when reviewed by a larger audience and finalized, could be used as part of IT procurement cycles to help DoD better evaluate risk.

As long as this is sensible, the questions are phrased to allow expository answers, and the benefit derived is commensurate with the cost of vendors completing it, this is one way for DoD both to know what they are getting and to put vendors on notice that quality and security-worthiness has become a purchasing criteria for DoD. There also needs to be some way for vendors to complete these questions so they are not repeating the same questionnaire for the same product (or subsequent releases of it) needlessly.

A Research and Development Program on Assurance is Needed

RECOMMENDATION: DoD should establish and fund a comprehensive Science and Technology Strategy and research programs to advance the state-of-the-art in vulnerability detection and mitigation within software and hardware.

NSA has established a Center for Assured Software, which is currently working a number of these issues. NSA should be resourced to act as the Executive Agent to manage the R&D program through the Center for Assured Software and to coordinate across the community to provide vulnerability detection and mitigation support to national security acquisition programs.

The goals of the classified and unclassified research and development investments in assurance should be to develop the technology to effectively take accidental vulnerabilities out of systems development and to improve Trusted Computing Group technologies in order to bound the risks presented by intentionally planted software. This program should monitor what markets are delivering, identify gaps between what the market is delivering and what DoD needs, and fill this gap.

Among the research programs that should be pursued are:

- Verification and model checking along with other forms of testing and static analysis.
- How hardware based security mechanisms can be used to improve software assurance.
- Development of high assurance components such as is accomplished within the Multiple Independent Levels of Security (MILS) program and the High Assurance Platform (HAP), and how to leverage the use of such high assurance components within system design.

Modest government unclassified research investments in model checking on source code are producing very valuable results already in broad use, and since these techniques still have substantial room for progress, continued and amplified investments in these tools seems prudent. In addition, since current model checking tools scale relatively well and emerging automated techniques for extracting immediate representations of binaries now provide enough fidelity to facilitate model checking, new and increased investments in scaling of immediate representation extraction for model checking also seem prudent.

Automation of reverse engineering may also lend itself to more powerful analysis techniques, such as theorem proving, that are currently less scalable but more effective in detecting hidden functionality in the very small software systems that are currently tractable. Theorem proving has been exceptionally valuable in the design of microprocessors with tens of millions of transistors, but has not scaled well in design and verification of software systems. However, theorem proving remains one of the few techniques capable of proving the absence of certain forms of hidden functionality. In this context, modest investments toward improving scalability of theorem proving seem worthwhile. Although successes have been harder to win, this area is not without recent successes of unclassified research investments, such as Proof Carrying Code and A Computational Logic for Applicative Common Lisp (ACL2) proofs of separation kernels.

The Task Force proposes unclassified investments in research and development of new and improved technology in three areas:

- Improving the scaling and improving the breadth and depth of accidental vulnerabilities found through model-checking-based source code analysis tools.
- Development of model checking on binaries through improvement of intermediate representation extraction techniques and their scalability.
- Scaling theorem proving for software systems.

The goal of such unclassified investments should be to develop technology that can reliably and effectively take accidental vulnerabilities out of the system. This would reduce the opportunity for adversaries to intentionally introduce vulnerabilities that masquerade as non-attributable accidental vulnerabilities. This, in turn, would force U.S. adversaries to more deeply and carefully hide malicious functionality, at greater expense, and possibly requiring a larger talent base more easily detected through intelligence means.

Classified Investments

Growing current investment in model checking of source code and investing in research toward model checking of binary code and scalable theorem proving of binary software all seem likely to provide help of a value that is many times the level of investment. However, none of these techniques will be able to resolve sufficiently the problem of reliably detecting well hidden malicious functionality.

For this reason, it seems important to not only make the above unclassified investments, which will provide the greatest relief, but to make these investments in parallel with investments in classified applications of such knowledge.

The National Agenda for Software Assurance

RECOMMENDATION: The DoD should advance the issue of software assurance and globalization on the national agenda as part of a coordinated effort to reduce national cyber risk.

The mandate of this Task Force was to address the systems assurance problem from the DoD perspective. DoD is addressing this issue aggressively because of the importance of maintaining mission-critical capabilities, but must be cognizant of the importance and benefits of advancing the issue of software assurance beyond the DoD realm. In reality, DoDs mission is inexorably linked to and reliant upon the work of related government agencies from the Intelligence Community to Homeland Security, and to companies that provide U.S. critical infrastructure.

Accordingly, this report should be read in the context of the even broader cyber risk that faces DoD, the U.S. Government generally, and the private sector. This risk emanates from two key factors: first, the national dependency on interconnected IT systems in public and private institutions for the Nation's national security, public safety and economic prosperity; and, second, the widespread vulnerability of these systems to attackers. A nation-state adversary cares only about fulfilling its mission, not about what is attacked, whether it is the DoD warfighting capability or the Hoover Dam. A holistic, coordinated approach to address the Nation's cyber risk must be a high priority for the national agenda.

One specific area that warrants DoD priority engagement of key government and private stakeholders is the supply chain risk to national security systems, a risk which has been exacerbated by globalization. This issue has been the focus of the Committee on National Security Systems' (CNSS) Global Information Technology Working Group, and its recent report, "Framework for Lifecycle Risk Mitigation for National Security Systems in the Era of Globalization." This report was endorsed unanimously by the 24 agencies with national security systems that comprise the CNSS. DoD should work closely with DHS to ensure that the U.S. Government expeditiously resources and implements the recommendations of this report.

RECOMMENDATION: DoD should work with its industry partners and commercial industry to improve the assuredness of COTS software.

Industry development processes and product evaluations (especially Common Criteria) should be brought into better alignment. Such coordination should enable product developers to produce the relevant product information evaluators need. Common Criteria should be considered in the evaluation of COTS products. Industry should employ modern secure development practices, such as threat modeling, use of vulnerability analysis tools, and focused security testing. Such modifications to industry development practices and evaluation processes will give integrators and end users products with better defined assurance properties.

APPENDIX

Appendix A.
Terms of Reference

THE UNDER SECRETARY OF DEFENSE
3010 DEFENSE PENTAGON
WASHINGTON, DC 20301-3010

OCT - 5 2005

ACQUISITION
TECHNOLOGY
AND LOGISTICS

MEMORANDUM FOR CHAIRMAN, DEFENSE SCIENCE BOARD

Subject: Terms of Reference – Defense Science Board Task Force on Mission Impact of Foreign Influence on DoD Software

You are requested to form a Defense Science Board Task Force to assess the risk that DoD runs as a result of foreign influence on its software and to suggest technology and other measures to mitigate that risk.

The DSB recently completed a study on national security implications of the migration of semiconductor manufacturing and design capability to foreign countries. The microelectronic task force offers that any strategy to mitigate concerns in semiconductor manufacturing would be of limited utility without a comparable strategy for dealing with the corresponding software challenge.

It is clear that the DoD's dependence on software is very high. Less well understood is the extent to which foreign influenced software is embedded within systems critical to our mission—both *de jure* and *de facto*. Nor do we understand fully the ability of DoD to perform its mission in an environment where it cannot depend on its software and where a portion of that software may have been intentionally compromised.

The Task Force should characterize our dependence on foreign-influenced software and assess the balance of risks associated with the findings; and assess the implications on DoD of a foreign entity being able to compromise our software-based systems throughout their full life cycle—design, development, production, testing, and maintenance, including attack vectors not necessarily unique to foreign suppliers. The Task Force should develop a strategy to effectively address these insults if warranted. The Task Force should become familiar with the microelectronic study and identify areas where the software and microelectronic strategies and associated recommendations should be coordinated.

While investigating these concerns, the Task Force will want to address the following questions:

a. What are the root causes associated with our current dependence on software that has foreign provenance? Are there policies or technology investments that DoD, either alone or in conjunction with other US government agencies, can pursue which will increase both the trustworthiness of software and our confidence in it?

b. What is our current understanding of adversary technical and operational capabilities and intentions to take advantage of these access opportunities? What requirements should be levied on the Intelligence Community to better enable us to characterize the threat?

c. In the event that suggested remedies do not "scale up," what criteria might be used to identify those software components that require the highest levels of trustworthiness? What is our current ability to evaluate the trustworthiness of a component/system against putative threats? What investments should be made to increase our confidence in the trustworthiness of a component/system? Once a component/system is deemed trustworthy enough, how do we ensure that this level of trust is maintained, or promptly noticed in the breach?

d. Which Defense and Intelligence Community organizations are key stakeholders for software assurance and what are their respective capabilities? What are the current research, technology, cadre, policy and legal challenges that must be addressed?

The Study will be co-sponsored by me as the Under Secretary of Defense (Acquisition, Technology, and Logistics), the Assistant Secretary of Defense (Networks and Information Integration), and the Commander, STRATCOM. Dr. Bob Lucky will chair the Task Force. Mr. Robert Lentz, OASD (NII), will serve as Executive Secretary and CDR Cliff Phillips will serve as the Defense Science Board Secretariat representative. It is envisioned that this study will require access to sensitive compartmented information.

The Task Force will operate in accordance with the provisions of P.L. 92-463, the "Federal Advisory Committee Act," and DoD Directive 5105.4, the "DoD Federal Advisory Committee Management Program." It is not anticipated that this Task Force will need to go into any "particular matters" within the meaning of Section 208 of Title 18, United States Code, nor will it cause any member to be placed in the position of action as a procurement official.

Kenneth J. Krieg

Appendix B.
Task Force Membership

CHAIRMAN

Dr. Robert Lucky — Telcordia Technologies

EXECUTIVE SECRETARY

Mr. Robert Lentz — OASD (NII)

MEMBERS

Mr. Scott Charney	Microsoft
Ms. Mary Ann Davidson	Oracle
Mr. James Gosler	Sandia National Laboratories
Mr. Paul Hoeper	Private Consultant
Dr. Joe Markowitz	Private Consultant
Mr. Richard Pethia	SEI, Carnegie Mellon University
Mr. Andy Purdy	Private Consultant
Mr. Steve Schanzer	S2 and Associates, Inc.
Dr. Trey Smith	SAIC
Hon. John Stenbit	Private Consultant
Mr. Brian Witten	Symantec
Mr. Larry Wright	Private Consultant

GOVERNMENT ADVISORS

Mr. William Dawson	CIA
Mr. Richard Hale	DISA
Mr. Craig Hasson	NSA
Mr. Larry Huffman	DISA
Mr. Joe Jarzombek	DHS
Mr. Mitch Komaroff	OASD (NII)
Mr. Phil McConnell	CIA
Mr. Wadiah Mikhail	OSD/IP

OBSERVERS

Dr. Greg Larsen	IDA
Mr. Bill Neugent	MITRE
Mr. Robert Reynolds	IDA

MILITARY ASSISTANT

CDR Cliff Phillips	U.S. Navy, Defense Science Board

SUPPORT STAFF

Ms. Michelle Ashley	SAIC
Ms. Diana Conty	SAIC
Mr. Jonathan Hamblin	SAIC

Appendix C.

Descriptions of Example Software-Intensive Programs in DoD

Blue Force Tracking (BFT)

Blue Force Tracking is an automated, friendly-force reporting system. Its use in Operation Enduring Freedom (OEF) and Operation Iraqi Freedom (OIF) demonstrated the value of a digitized army and provided the operational evidence of the value of Net-Centric operations. The combination of Blue Force Tracking with the Army's platform-level battle command system, FBCB2, generates and distributes a common view of the battle space at the tactical and operational levels. In OEF/OIF, Force XXI Battle Command, Brigade-and-Below/Blue Force Tracking (FBCB2/BFT) enabled tactical units to perform in conditions that would have significantly degraded operations in the first Gulf War. Combat deaths from friendly fire were dramatically reduced relative to the earlier conflict.[1] Furthermore, in some critical situations units were able to maneuver based only on FBCB2/BFT reports and map displays.

In 2002, the Army G6/CIO and Program Executive Office for Command, Control, and Communications - Tactical (PEO C3T) initiated an effort to deploy Blue Force Tracking to the U.S. Army, U.S. Marine Corps, and U.K. Army units that would fight in OEF/OIF. Drawing on previous experience with a Ku-band architecture fielded for Balkan peacekeeping operations, the office of the G6 designed an architecture that uses commercial L-band satellite links for beyond-line-of-sight communication. There are presently more than 20,000 platforms equipped with FBCB2/BFT in Iraq and Afghanistan.

This program has used both commercial and military devices as hardware. FBCB2/BFT uses COTS products and components from the Defense Information Infrastructure Common Operating Environment as appropriate, keeping in perspective the intended "small computer" characteristic of the FBCB2 design. Blue Force Tracking was delivered to troops in Afghanistan and Iraq using Intel computers. The software code is a combination of Linux, GOTS, and COTS (including Oracle-based software.) Software is upgraded and maintained via physical distribution of hard drives with masters. The masters are delivered to operational units, which then clone them for installation in platforms.

- **Information Safeguards:**[2] FBCB2 processes, sends, receives, displays and stores information classified up to SECRET level. (Note: BFT is not accredited to

---

[1] In the first Gulf War, friendly fire caused 25.6% of combat deaths. Through the end of "major combat operations" in the Second Gulf War (May 2, 2003), friendly fire caused only 6.5% of the combat deaths. Sources: Oscar Avilla "U.S. Won with Few Casualties," Chicago Tribune, May 3, 2003 and Peter Pae "'Friendly Fire' Still a Problem," Los Angeles Times, May 16 2003.

[2] From Test & Evaluation Master Plan Part 1b (23 Aug 04) at www.hqda.army.mil/tema/TEMP%20101/PART1/PART%201B%20FBCB2%2023%20Aug%2004.doc

Appendix

operate at the SECRET level. The system runs a Type II encryption. Training and Doctrine Command Systems Manager (TSM) FBCB2 has accepted this limitation for the present based on operational need while efforts continue towards a Type I implementation and accreditation).

No software source selection decisions on FBCB2/BFT have been made since 1995. Supplier assurance was not an explicit requirement of the RFP process. Likewise, Foreign Ownership, Control or Influence of COTS component suppliers was not a criterion in the selection process. The Program Manager has, however, identified and reviewed the software security and configuration management policies used by the lead systems integrator. FBCB2/BFT has a current STAR (System Threat Assessment Report) dated 7 March 2005. The formal policy for handling source and executable code is in final draft as part of an amendment to the 30 Nov 2005 Program Protection Plan.

FBCB2/BFT has succeeded in delivering a Net-Centric capability to warfighters that has dramatically increased mission effectiveness. The system incorporates commercial, COTS, GOTS and Open Source IT hardware and software. The overall approach to system security was developed under existing policies. Because information rides on commercial links, encryption and authentication mechanisms have been a critical element in information assurance and in protecting the FBCB2/BFT network from intrusion. FBCB2/BFT has been operationally effective in war.

While the Task Force sees no evidence that system effectiveness was degraded by foreign influence on its component hardware or software, it must be noted that the United States was facing an unsophisticated adversary, which is not the main focus of this paper. BFT Project Management Officer (PMO) made no effort to identify the pedigree of system components of BFT through CI support to source selection. It has been significantly aided in this decision by the lag-time since last source selection for COTS integration, which may pre-date attention to the Systems Assurance Problem. Future source selection of COTS integration decision will in all likelihood be supported by intelligence support as in FCS described below which, while far from ideal, are stronger than used in BFT.

Future Combat Systems

The Task Force received a summary briefing on FCS Software Assurance from Army and contractor personnel on 17 October 2006. Subsequently, on 14 November 2006, two members of the Task Force[3] received a thorough briefing on FCS Information Assurance from the programs information assurance deputy.

The FCS program will develop and field a mobile, self-healing, "plug and play" network that will link together a family of light, mobile and survivable manned vehicles; unattended aerial vehicles; unattended ground vehicles; highly capable sensors and advanced lethal munitions within a Brigade Combat Team.

---

[3] Trey Smith and Page Hoeper

Each FCS Brigade Combat Team will be networked through the Army Battle Command System. Vertical networks will link command echelons while horizontal connections will tie together mounted, dismounted and airborne combat units. The network architecture will include every stationary and moving platform in the battle space. The Army Battle Command System will be interoperable with theater, joint and combined command and control systems. FCS network connections will include the JWICS (Joint Worldwide Intelligence Communication System), the SIPRNET, and the NIPRNET/Internet.

The IT hardware and software within FCS includes COTS, GOTS, Open Source, Proprietary and Military Unique components. For software in particular, about 27 million source lines of code are COTS or open source. This constitutes over 42% of the total delivered executable source lines of code.

The System of Systems Common Operating Environment (SOSCOE) and the Integrated Computer System/Operating System (ICS/OS) rely predominately COTS and Open Source software. The ICS/OS is almost 99% COTS/OS. The SOSCOE, essentially the "middleware" for FCS, is almost 80% COTS/OS.

As with many other DoD software applications, the use of COTS/GOTS/OSS is essential to achieving required capabilities within time and budget constraints. In some cases there may be no other practical way to deliver the required capabilities. Furthermore, COTS/GOTS/OSS can be highly secure. The FCS operating system, part of the ICS, employs LynxOS-SE, a security enhanced real time operating system from LynuxWorks.

The FCS program office has assessed that there is a low to moderate risk that malicious code could be inserted into the FCS Master Software Baseline and exploited. The program has identified the following counter measures to reduce that risk to an acceptable level:

- Establish "software pedigree."
- Incorporate malicious code detection tools into the software development process.
- Execute defined software configuration management processes.
- Detect and remove "dead" code.
- Use secure code development processes.

Regarding the first bullet above, the Future Combat System Chief Information Officer (FCS CIO) provided initial guidance (pending Program approval) on the use of foreign code for FCS. Key points include:

- Government assumption that the profit motive will assure clean code in "shrink wrapped" (consumer) software.
- Code certification for software that is specifically developed for a known government use.
- Acceptance of foreign software for use in areas that are not critical to the performance of SOSCOE.

- Purchase of foreign software via a "blind buy," in which the seller has no knowledge of who the buyer is or how the product will be used.
- Approval of Open Source Software only under stringent conditions.

Recognizing that the above points are in draft, the Task Force would urge caution in accepting the proposition that the profit motive and intent to sell software to a mass market will induce sufficient testing to produce clean code. The Task Force is similarly skeptical of the proposition that very much FCS software can be acquired through blind buys, where the seller would not be aware of the ultimate user or the intended placement or use of the software.. And the Task Force lacks confidence in current tools for detecting malicious code.

Malicious code is a key concern of the FCS Program. According to the Deputy for Information Assurance, the Program recognizes that current automated tools cannot detect all malicious code and inherent vulnerabilities. Current policy is line-by-line code inspection, which the program recognizes is not feasible. At present, the effort to detect malicious code depends on detection of anomalous behavior.

The FCS Program recognizes that where malicious code is concerned, prevention during development and maintenance is the best policy. Prevention includes careful vetting of software pedigree, especially for the SOSCOE and the ICS/OS. Software maintenance will be secured through a Public Key Infrastructure (PKI) access control.

The FCS Program appears to have a robust and detailed information assurance plan. Software assurance is a component of this plan. The Task Force recommends that DoD NII meet with the FCS information assurance deputy to exchange views on how best to assure that FCS is not compromised by foreign influence on its software and overall information technology.

Near Real-Time, Sensor-to-Weapon Considerations

Software assurance is more mature and well developed in DoD weapons systems, particularly those with direct, near real-time linkage from the sensor to the weapon control system. At the same time, the size, complexity and potential vulnerability of the networks that connect sensors to weapons are increasing. Systems are beginning to migrate toward industry standard and Internet protocol (IP) technologies. Evidence gathered during this study shows that sensor-to-weapon networks range from secure, special purpose data and communications links, such as the USAF F-22 program, to networks that rely on industry standard protocols and links, such as the Army's FCS. The constantly increasing capabilities offered by industry standard protocols drives an irresistible tendency to replace relatively secure special-purpose communications protocol stacks with the general purpose Internet protocol (IP) stack.

Given this tendency, assurance practices, such as those embedded in the FCS program must be well supported and stringently implemented. The program has established an FCS Software Assurance Focus Group to define policies and direction to ensure "due diligence" in identifying, assessing and managing software information assurance risks associated with the development and acquisition of software for the FCS master software baseline. The Focus Group consists of a lead system integrator, government, and subject matter experts. The Group has established a four-pronged approach centered on software acquisition processes, software development processes, software testing requirements and software configuration management execution. The Focus Group maintains proactive processes to identify and manage future risks. The practices and processes employed in the FCS program may indeed offer protection from less sophisticated, weakly resourced adversaries, but do not appear sufficient to protect program capabilities from damage by a sophisticated nation-state adversary intent on damaging FCS effectiveness.

F-22 Program Overview

The F-22 Raptor is a fifth generation stealth fighter aircraft. Raptor formally entered U.S. Air Force service in December 2005 as the F-22A. Its advanced capabilities depend on sophisticated software and hardware to process and analyze signals and data. Given the central and critical nature of these systems, the F-22 initiated development with defined security requirements intended to eliminate the possibility of malicious software. The program developed a security architecture including firewalls to protect key components and then established criticality classifications for each Computer Software Configuration Item (CSCI). The software development processes were tailored to match the criticality of the items and the program.

The F-22 computing system architecture is segregated into two domains based on the criticality of the data. There is a non-trusted or non-sensitive environment that handles non-critical data and operations that do not require stringent controls and protection. There is a Trusted Computing Base (TCB) domain that contains software that processes sensitive data or algorithms. It comprises all the avionics software and data that is sensitive or critical to platform and mission performance. The TCB was built in compliance with the TCB specifications. The TCB domain protects all data through the means of a firewall provided by software and hardware specifically designed for the protection of the classified information. All TCB components performing the firewall functionality are classified as Protection-Critical Trusted and are subjected to more stringent development processes. Any classified information that is transferred outside the TCB is encrypted. Protections within the TCB also ensure that data is only shared between software applications that have a predetermined (system design) "need to know" via defined interfaces. The build process creates access control rights tables through the Privilege Control Table Toolkit, which is also created using protection critical processes.

F-22 Software assurance is maintained through security policies which are defined and enforced through the F-22 Weapon System Certification and Accreditation (C&A) Plan (WS SCAP) and Air Vehicle C&A Plan. Security Functional and Assurance Requirements are based on legacy processes including NISPOM, Joint Air Force/Army/Navy (JAFAN) 6/3, 6/4, 6/9, and DoD 8500.1, Department of Defense Instruction (DODI) 8500.2

Software Development

The F-22 Software is developed via the F-22 Weapon System Software Development Plan (WS SDP). It includes specific process requirements for trusted and Protection-Critical Trusted Software. Compliance with assurance requirements are substantiated via process verification and documentation. Specific requirements prohibit software developed by any foreign national that is part of the trusted software load. Above and beyond all this, protection-critical trusted software requires a code review independent of the developing organization to ensure that all such code meets its functional requirements and contains no unwanted code and accesses only data needed to meet its functional requirements.

The only software developed by foreign partners that is part of the F-22 Air Vehicle is the embedded software for the Heads-Up Display (HUD). This software displays symbology and resides outside the TCB (no access to classified information) and does not have access to critical information.

Test and evaluation

Software Tools used to develop, integrate and test the F-22 OFP are part of the System/Software Engineering Environment (S/SEE), which governs the acceptance of software development tools. All tools go through a Product Security Evaluation (PSE) on isolated computer networks prior to being accepted for use on the F-22 program. These tools test for unexpected behaviors, events and data flow and analyze source objects for non approved components (e.g. shareware).

Future Plans

Modernization of the F-22 avionics envisions the incorporation of commercial software products. In fact, current F-22 modernization efforts are incorporating commercial operating systems (OS) such as the Green Hills Integrity 178B. This software was developed to standards that go beyond current F-22 requirements in order to meet National Security Telecommunications and Information Systems Security Policy (NSTISSP) 11. This software also meets Separation Kernel Protection Profile (SKPP) requirements. It was developed by NSA and industry and approved through NSA. It is now being evaluated to Evaluation Assurance Level 6+ (EAL 6+) by the Air Force Research Laboratory.

Conclusions

Risk from malicious software was mitigated by the F-22 program based on conscious decisions since program initiation. These decisions cover all sources of malicious software from both foreign sources and U.S. citizens. Established facility, personnel and process security requirements help to ensure protections from malicious software. The F-22 program stands at one extreme of the software development and assurance spectrum as currently executed in DoD. However, the cost and investment required to develop systems in a similar fashion would be prohibitive for many DoD programs. The F-22 appears to be at the high end of weapons programs when measured against current efforts for secure software development of its most critical systems.

Appendix D.

Task Force Findings and Recommendations

Findings

- The software industry, as well as the software talent base, is becoming increasingly global, and this trend appears to be irreversible.

- DoD has become increasingly dependent for mission-critical functionality upon highly interconnected, globally-sourced, information technology of dramatically varying quality, reliability and trustworthiness.

- Adversaries' use of low-level cyber attack techniques to exploit weak information assurance controls and vulnerable software has led to successful attacks upon sensitive but unclassified (SBU) systems and networks of DoD and the defense and commercial industrial base. This has occurred notwithstanding extensive DoD efforts in security and information assurance.

- The enormous functionality and complexity of IT makes it easy to exploit and hard to defend, resulting in a target that can be expected to be exploited by sophisticated nation-state adversaries.

- Nation-state adversaries are able to employ a full spectrum of offensive intelligence trade-craft, including attacks that subvert the supply chain, to damage or defeat mission-critical systems. The risk from such supply chain exploits can only increase as larger portions of the supply chain become more accessible to the adversary through global sourcing.

- Nation-state adversaries' penetration of sensitive-but-unclassified systems and networks could allow them to steal system information or tamper with system artifacts, enabling them to target existing and future DoD Systems.

- DoDs defensive strategies and techniques remain inadequately informed of the sophisticated capabilities of nation-state adversaries to exploit globally sourced, ubiquitously interconnected, COTS HW/SW within DoD Critical Systems, and the potential consequences of system subversion.

- The Intelligence Community (IC) does not deliver timely, actionable intelligence regarding the intents and capabilities of nation-state adversaries to attack and subvert DoD systems and networks through supply chain exploitations, or through other sophisticated techniques.

- The risk of undetected subversion of custom software is considerably greater than the corresponding risk for COTS.

- Software deployed across DoD continues to contain numerous vulnerabilities and weak information security design characteristics. DoD and its industry partners spend considerable resources on patch management, while gaining only limited improvement in defensive posture.

- The primary processes relied upon by DoD for evaluation of the assurance of commercial products, FIPS and NIAP Common Criteria Evaluation Process, do not address software vulnerabilities except for higher assurance levels (if at all). Moreover, these processes do not scale to the volume of software critical to the DoD mission.

- The Task Force identified considerable variation in the extent to which the Systems Assurance Problem is impacting next-generation DoD systems. That impact ranges from extensive with the introduction of inter-networked COTS and open source IT into the Army's Future Combat System (FCS) program, to only slight in the USAF F-22 program.

- DoD Critical Systems and networks remain vulnerable to the sophisticated capabilities of Nation-state adversaries. USD(AT&L) and ASD(NII)/DoD CIO, through the efforts of the DoD Software Assurance Tiger Team and the CNSS Globalization of IT Working Group (GITWG), have developed a comprehensive strategy and CONOPS for addressing the systems assurance problem within DoD and an approach for the Federal Government at large.

- DoD does not consistently or adequately analyze and incorporate into its acquisition decisions the supply chain threat information that is available.

- The growing complexity of the microelectronics and software within its critical systems and networks makes DoDs current test and evaluation capabilities unequal to the task of discovering unintentional vulnerabilities, let alone malicious constructs.

- The problem of detecting vulnerabilities is deeply complex and there is no silver bullet on the horizon.

- The academic curriculum and courseware in computer science does not adequately teach software developers to design, develop and test with a defensive mindset, nor is there adequate training in assurance and security techniques.

Recommendations

- DoD should continue to procure from, encourage and leverage the largest possible global competitive market place consistent with national security.

- The IC should be tasked to collect and disseminate intelligence regarding the intents and capabilities of adversaries, particularly nation-state adversaries, to attack and subvert DoD systems and networks through supply chain exploitations, or through other sophisticated techniques.

- DoD should increase relevant knowledge and awareness among its cyber-defense and acquisition communities of the capabilities and intents of nation-state adversaries.

- DoD should allocate assurance resources among acquisition programs at the architecture level based upon mission impact of system failure.

- DoD should implement a scalable supplier assurance process to assure that critical suppliers are trustworthy. DoD should coordinate with the DNI to leverage the all-source threat assessment capability resident in the DNIs Intelligence Community Acquisition Risk Center (CARC), possibly shaping this into the foundation of a national supply chain threat assessment capability.

- DoD acquisition processes should require that products possess assurance matching the criticality of the function delivered.

- DoD should require that all custom code written for systems deemed critical be developed by cleared U.S. citizens.

- DoD should provide incentives to industry to produce higher quality code.

- DoD should establish and fund a comprehensive Science and Technology Strategy and research programs to advance the state-of-the-art in vulnerability detection and mitigation within software and hardware.

- The DoD should advance the issue of software assurance and globalization on the national agenda as part of a coordinated effort to reduce national cyber risk.

- DoD should work with its industry partners and commercial industry to improve the assuredness of COTS software.

Appendix E.

Briefings Presented to the Task Force

NAME	TOPIC
23 FEBRUARY 2006	
Mr. Jeffrey Green, DoD General Counsel	Task Force Ethics
Dr. James Lewis, CSIS	Foreign Influence on DoD Software
Mr. Taylor Scott, DIA	Chinese Cyber Threat (Classified)
Mr. Donovan Lewis, DIA	IT Globalization and DoD Network Security (FOUO)
Mr. Robert Reynolds, IDA	Software Assurance Threats (Classified)
24 FEBRUARY 2006	
Mr. Kris Britton, NSA	Software Assurance and Vulnerability Discovery (Classified)
Dr. Catherine Mann, Institute for International Economics	Globalization of Information Technology
Mr. Phil McConnell, CIA	Risk Management of the Intelligence Community Acquisition Risk Center (CARC) (FOUO)
Mr. Mitchell Komaroff, OASD (NII)	DoD Software Assurance Tiger Team Strategy
Mr. William Archey, American Electronics Association	Trends in Globalization and Offshoring
22 MARCH 2006	
Mr. Jeffrey Payne, Cigital, Inc.	Software Security
Dr. Brian Chess, Fortify Software, Inc.	Software Security and Validation (Static Analysis)
Ms. Susan Lee, Johns Hopkins University	Countering Untrustworthy Software
Dr. William Howard, Private Consultant	DSB Task Force on High Performance Microchip Supply

NAME	TOPIC
18 April 2006	
Mr. David Grawrock, Intel Corporation	Trusted Platforms
Mr. Grant Wagner, NSA	Trusted Platforms and Software Security
Mr. Joe Jarzombek, DHS	Overview of the DHS Software Assurance Program
16 May 2006	
Mr. Robert Alexander, U.S. Army Mr. Tom Plavcan, U.S. Army Mr. Jack Saylor, Information Assurance Security Office	Blue Force Tracker Program
Ms. Audrey Dale, NSA	Common Criteria Evaluation and Validation Scheme (CCEVS)
Mr. John Goodenough, Software Engineering Institute Mr. Bill Scherlis, Carnegie Mellon University	Software Assurance Opportunities
16 June 2006	
Dr. Ben Calloni, Lockheed Martin	Multiple Independent Layers of Security (MILS) Program
Mr. Frank Hecker, Mozilla Foundation	Open Source Software
29 August 2006	
Mr. Brian Rianhard, Lockheed Martin	F-22 Software
17 October 2006	
Mr. Rand Blazer, SAP America, Inc.	Perspective on Software in a Global Environment
Mr. Edgar Dalrymple, U.S. Army Mr. Robert Myers, SAIC	Future Combat System (FCS) Software

Appendix F.

References

[CMMI Product Team 2006]
CMMI Product Team; CMMI for Development, Version 1.2, (CMU-SEI-2006-TR-008) Pittsburgh, PA: Software Engineering Institute, Carnegie Mellon University, 2006

[Cooper 2002]
Cooper, J.; Fisher, M. Software Acquisition Capability Maturity Model (SA-CMM), Version 1.03, (CMU-SEI-2002-TR-010) Pittsburgh PA: Software Engineering Institute, Carnegie Mellon University, 2002

[Curtis 2003]
Curtis, B.; Hefley, W.; Miller, S. Overview of the People Capability Maturity Model (P-CMM), (CMU-SEI-2003-MM-001) Pittsburgh, PA: Software Engineering Institute, Carnegie Mellon University, 2003

[Dion 1993]
Dion, R. "Process Improvement and the Corporate Balance Sheet." IEEE Software, 10, 4 July 1993): 28-35.

[DoD 1996]
Department of Defense. DoD Guide to Integrated Product and Process Development (Version 1.0) Washington, DC: Office of the Under Secretary of Defense (Acquisition and Technology), February 5, 1996.

[EIA 1998]
Electronic Industries Association. Systems Engineering Capability Maturity Model (EIA/IS-731). Washington, DC: Electronic Industries Association, 1998.

[Goldenson 2003]
Goldenson, D. & Gibson, D. Demonstrating the Impact and Benefits of CMMI: An Update and Preliminary Results, (CMU-SEI-2003-SR-009) Pittsburgh, PA: Software Engineering Institute, Carnegie Mellon University, 2003.

[Herbsleb 1994]
Herbsleb, J.; Carleton, A.; Rozum, J; Siegel, J.; & Zubrow, D. Benefits of CMM-Based Software Improvement: Initial Results (CMU-SEI-94-TR-13) Pittsburgh, PA: Software Engineering Institute, Carnegie Mellon University, 1994.

[Howard 2002] Strsafe.h: Safer String Handling in C
"http://msdn.microsoft.com/library/en-us/dnsecure/html/strsafe.asp"

[Howard and Lipner 2005] Howard, M. and Lipner, S. The Security Development Lifecycle. Microsoft Press 2005.

[ISS 2007] IBM Internet Security Systems, "X-Force 2006 Trend Statistics", January 2007. http://www.iss.net/documents/whitepapers/X_Force_Exec_Brief.pdf

[Lawrence 2006] Kenneth L. Lawrence (MAJ, USA) and Glen Foster. "Clarifying the DoD Vulnerability Management Program." August 30, 2006. Joint Task Force-GNO Technical Bulletin 06-006

[Lemos 2006] Robert Lemos, "Security flaws on the rise, questions remain" SecurityFocus, 2006-01-05 http://www.securityfocus.com/news/11367

[Lipke 1992]
Lipke, W. & Butler, K. "Software Process Improvement: A Success Story." Crosstalk 5, 9 (September 1992): 29-39.

[Lovell 2005] Repel Attacks on Your Code with the Visual Studio 2005 Safe C and C++ Libraries http://msdn2.microsoft.com/en-us/library/8ef0s5kh(VS.80).aspx

[Messmer 2007] Messmer, Ellen, "Software vulnerabilities spiked 39% in 2006", Network World, 01/30/07, http://www.networkworld.com/news/2007/013007-ibm-security-report.html

[Paulk 1993]
Paulk, M.; Curtis, B.; Chrissis, M.; Weber, C. Capability Maturity Model for Software (Version 1.1) (CMU-SEI-93-TR-024) Pittsburgh, PA: Software Engineering Institute, Carnegie Mellon University, 1993.

[Sims 2005]
Sims, Jennifer E. and Burton Gerber, eds. Transforming U.S. Intelligence. Washington: Georgetown University Press, 2005

[Wohlwend 1993]
Wohlwend, J. & Rosenbaum, S. "Software Improvements in an International company" 212-220. Proceedings of the International Conference on Software Engineering, Baltimore, Maryland, May 17-21, 1993. Los Alamitos, CA: IEEE Computer Society Press, 1993

Appendix G.
Abbreviations & Acronyms

ACL2	A Computational Logic for Applicative Common Lisp
ACM	Association of Computing Machinery
ARA	Affine Relational Analysis
ASD(NII)	Assistant Secretary of Defense for Networks and Information Integration
ASI	Aggregate Structure Identification
ASIC	Application-Specific Integrated Circuit
BFT	Blue Force Tracking
C&A	Certification and Accreditation
C2	command and control
CAA	Common Assurance Assessment
CARC	Community Acquisition Risk Center
CC	Common Criteria
CMM	Capability Maturity Model
CMMI	Capability Maturity Model Integration
CNSS	Committee on National Security Systems
CONOPS	Concept of Operations
COTS	Commercial-off-the-shelf
CSCI	Computer Software Configuration Item
CSIS	Center for Strategic and International Studies
DHS	Department of Homeland Security
DIACAP	DoD IA Certification and Accreditation Process
DNI	Director of National Intelligence
DoD	Department of Defense
DoD CIO	Department of Defense Chief Information Officer
DODI	Department of Defense Instruction
EAL	Evaluation Assurance Level
FBCB2	Force XXI Battle Command, Brigade-and-Below
FCS	Future Combat Systems

FCS CIO	Future Combat Systems Chief Information Officer
FIPS	Federal Information Processing Standards
FTE	full time equivalent
GAO	Government Accountability Office
GITWG	Globalization of IT Working Group
GOTS	Government-off-the-shelf
HAP	High Assurance Platform
HUD	Heads-Up Display
HW	hardware
I/O	input/output
IA	Information Assurance
IAV	Information Assurance Vulnerability
IAVA	Information Assurance Vulnerability Alert
IAVB	Information Assurance Vulnerability Bulletin
IAVM	Information Assurance Vulnerability Management
IAVT	Information Assurance Technical Advisory
IBM	International Business Machines
IC	Intelligence Community
ICARC	Intelligence Community Acquisition Risk Center
ICS	Integrated Computer System
IEEE	Institute of Electronics and Electrical Engineering
IP	Internet Protocol
ISC	Internet Systems Consortium
IT	Information Technology
ITAR	International Traffic in Arms Regulations
JAFAN	Joint Air Force/Army/Navy
JWICS	Joint Worldwide Intelligence Communication System
RFC	Request for Comments
MILS	Multi-Independent Levels of Security
NIAP	National Information Assurance Partnership
NIPRNET	Non-classified Internet Protocol Router Network
NISP	National Industrial Security Program

NISPOM	National Industrial Security Program Operating Manual
NIST	National Institute of Standards and Technology
NSA	National Security Agency
NSTISSP	National Security Telecommunications and Information Systems Security Policy
OEF	Operation Enduring Freedom
OIF	Operation Iraqi Freedom
OMB	Office of Management and Budget
OS	Operating System
OSS	Open Source Software
OUSD(AT&L)	Office of the Under Secretary of Defense for Acquisition, Technology and Logistics
PEO C3T	Program Executive Office for Command, Control, and Communications - Tactical
PKI	Public Key Infrastructure
PMO	Project Management Officer
PSE	Product Security Evaluation
R&D	research and development
RAM	Random-Access Memory
RFP	Request for Proposal
SA-CMM	Software Acquisition - Capability Maturity Model
SAP	Systems, Applications and Products
SBU	Sensitive But Unclassified
SEI	Software Engineering Institute
SIBS	Software Industrial Base Study
SIPRNET	Secret Internet Protocol Router Network
SKPP	Separation Kernel Protection Profile
SLOC	source lines of code
SOSCOE	System of Systems Common Operating Environment
STAR	System Threat Assessment Report
SW	software
TCB	Trusted Computing Base
TCG	Trusted Computing Group

TPM	Trusted Platform Module
TSM	Training and Doctrine Command Systems Manager
UAV	Unmanned Aerial Vehicle
U.K.	United Kingdom
U.S.	United States
USAF	United States Air Force
VSA	Value Set Analysis
WMD	Weapons of Mass Destruction
WS SCAP	Weapon System Certification and Accreditation Plan
WS SDP	Weapon System Software Development Plan
Y2K	Year 2000

www.ingramcontent.com/pod-product-compliance
Lightning Source LLC
Chambersburg PA
CBHW080706190526
45169CB00006B/2260